Unrevealed Milestones

in the

Iraqi National Nuclear Program

1981-1991

I0423209

By

Dhafir Selbi and Zuhair Al-Chalabi

Co-authored and edited by Dr. Imad Khadduri

ISBN-13: 978-1466418400

Cover design

Hadeel Al-Khozaie and Velinka Qutteineh

Dedication

To all Iraqi men and women who were committed to the Iraqi National Nuclear Program and who dearly sacrificed many of their working years in order to advance Iraq's scientific and technical achievements, especially during times of war.

To our families who had to patiently endure our almost continuous absence, and for their support, devotion and encouragement during troubled years.

Table of Contents

Preface

"*Sir* (addressing Saddam Hussain), *we will accomplish the mission by 1990*".

That is the potent promise that was offered on **April 1985** by some of the top executives in the Iraqi National Nuclear Program (INNP) to the late Saddam Hussain in a meeting that was held in a mobile caravan in Al-Radwanya, a heavily fortified site near Baghdad international airport.

The meeting was exclusively attended by Saddam Hussain, his deputy, his personal secretary and the five commissioners (members of the executive committee) of the Iraqi Atomic Energy Commission (IAEC).

Did that loaded promise impact the thinking and planning of Saddam Hussain that would eventually lead him to invade Kuwait?

This book sheds light on the most important milestones, some of them revealed here for the first time, during the ten years' progress, from 1981-1991, of the Iraqi National Nuclear Program, a covert and thorough Iraqi nuclear program that focused initially on several processes for uranium enrichment and then later turned to military objectives before its demise and destruction in 1991. Notably, this book sets the records straight in an unequivocal manner.

The defining moment for the program came exactly two years later, in **April 1987,** when the same members of the executive committee of the Iraqi Atomic Energy Commission who had made the solemn promise to Saddam Hussain now admitted to the rest of the members of the executive committee of the IAEC that the promised target date would not be met. This information is believed to have not been transmitted to Saddam Hussain at all. However, the consequences of the failure to fulfill the promise given on 1985 morphed the Iraqi National Nuclear Program into high gear, streamlined its administrative and research processes and led to the industrial scale adoption of the same enrichment process that was previously adopted in the American Manhattan Project in World War II.

This book also strives to put the program's record straight as it is recounted by participants who were in various leading positions in it as well as inputs and edits from other senior colleagues. It aims to debunk misleading accounts in two previously published books on this subject; i.e. "Saddam's Bombmaker: The Terrifying Inside Story of the Iraqi Nuclear and Biological Weapons Agenda" published in 2000, which provided foreign intelligence laden misinformation for the pretext for the invasion of Iraq, and "The Bomb in My Garden: The Secrets of Saddam's Nuclear Mastermind" written in 2004, which overtly and grossly overstated the role of its author in the program and was written to repay the debt of the authors' relocation to the U.S. by the CIA after the invasion of Iraq in 2003.

Chapter 1: Background on the Iraqi Nuclear Program

Introduction

For the duration of nearly a decade, from 1981 to 1991, Iraq had embarked on an extensive and covert scientific program which was referred to by the Iraqi authorities as the Iraqi National Nuclear Program (INNP). It was initially focused on investigating various uranium enrichment methods. Only during the last four years of its existence, did it also turn its attention to the feasibility of utilizing the application of enriched uranium for military use.

Since the occupation of Iraq in 2003, several books have been published by Iraqis who were directly involved in the Iraqi National Nuclear Program [1, 2, 3, 4], aside from two other books

---

[1] "**Iraq's Nuclear Mirage – Memoirs and Delusions**", by Dr. Imad Khadduri, in English (Springhead Publishers, Canada, ISBN 0-9733790-0-6), 2003, and in Arabic (Arab Scientific Publishers, Lebanon, ISBN 9953-29-974-9), 2004.

[2] "**Iraqi Weapons Mass Destruction (WMD): Fact and Fiction**" (Title in Arabic: "**The Last Confession – The Truth about the Iraqi Nuclear Program**"), by Dr. Jafar Dhia Jafar and Dr. Numan Al-Niaimi, in Arabic (Centre for Arab Unity Studies, Lebanon, ISBN 9953-450-99-4), 2005.

[3] "**Files from the Nuclear Program and the Military Industrialization**", in Arabic, by Dr. Basel Al-Saati (Arab Scientific Publishers, Lebanon, ISBN 9953-29-887-4), 2006.

11

that were written by former Iraqi scientists who now reside in the U.S.. However, the INNP community at large has major reasons to discredit both of these two tainted books[5, 6] as they are burdened with a heavy toll of foreign Intelligence presumptions and disinformation that has by now been manifestly demonstrated. In addition, there are extensive reports of the UN inspectors in its Iraq Nuclear Verification Office[7] and documentation presented to the UN by Iraq itself[8]

---

[4] **"The Strategy of the Nuclear Program in Iraq within the Context of Science and Technology Policies**", by Dr. Humam Abdul Khaliq and Dr. Abdulhaleem Al-Hijjaj, in Arabic (Centre for Arab Unity Studies, Lebanon, ISBN 978-9953-82-247-1), 2009.

[5] "**Saddam's Bombmaker**: The Daring Escape of the Man Who Built Iraq's Secret Weapon" by Khidhir Hamza and Jeff Stein, Touchstone Simon and Schuster Publishers, ISBN 0-684-87386-9), 2001.

[6] "**The Bomb in My Garden:** The Secrets of Saddam's Nuclear Mastermind" by Mahdi Obeidi and Kurt Pitzer, John Wiley and Sons, ISBN 0-471-67965-8), 2004.

[7] Fourth consolidated report of the Director General of the International Atomic Energy Agency under paragraph 16 of Security Council resolution 1051 (1996), in English 1997 (Report S/1997/779, http://www.iaea.org/OurWork/SV/Invo/reports/s_1997_779.pdf) which includes a comprehensive summary of the IAEA´s activities and achievements in Iraq between April 1991 and October 1997.

[8] "**Disarming Iraq**" (Title in Arabic: "**Disarming Iraq – Occupation instead of Inspection**"), by Dr. Hans Blix, in Arabic (Centre for Arab Unity Studies, Lebanon, ISBN 9953-82-010-4), 2005.

during the UN inspection era, which lasted from 1991 until the invasion of 2003. Yet, and despite these publications, there are major issues and revealing milestones which affected the path and the outcome of the INNP that have not yet been hitherto disclosed and some of these are now detailed in this document.

This book has been compiled by a group of senior members of the Iraqi National Nuclear Program who made major contributions to the INNP. Its content was coordinated by a member of the executive committee of the Iraqi Atomic Energy Commission during the nineteen eighties and who was heavily involved in the direction of the INNP during its last four years of existence.

It is important to point out at the outset that this book does not add to the content of Iraq's Full, Final and Complete Disclosure (FCCD) to the United Nations Special Commission (UNSCOM). It is rather intended to (i) unfold some of the most important embedded events that evolved in the management of the INNP, and (ii) the evolution of its work procedures. Both of these factors did seriously impact the course of the evolution of the program and the attitude to its policies at large.

It was intended that the authors of this book involve participants from different disciplines and who had held different upper level positions in the program with variant points of view on the path and the accomplishments of the INNP. Hence, the joint compilation and distillation of this book through the consensus of the participants is a strength that enhances the objectivity of the book that may not be found in the other books and documents on the INNP. The modest hope of this approach is that its content would be more objective in its presentation and appraisal of the INNP regarding both its accomplishments and its shortcomings.

A probable drawback to the contents of this book is the fact that almost all of the related important documents and archives were destroyed either through air bombardment during the armed hostilities or as a result of the post-war order by the Iraqi authorities for the scientist and participants to destroy all evidence of their work that was in their personal possession. Hence, the authors had to depend on their collective and interactive memory as reference and verification.

Furthermore, the authors wish to point out that some of the events and issues discussed in this book have been touched upon in the previous books on the INNP. The authors have strived to highlight and focus on the issues that they believe were not sufficiently covered previously as well as to provide crucial untold information on the INNP.

The authors wish to state unequivocally their highest degree of appreciation to the unique achievements of the INNP in totality which was the collective output of the INNP community who devoted the prime years of their lives and had selflessly sacrificed their social and personal family lives to the achievements of this program. Nevertheless, the authors maintain and feel that in return for their devotion and sacrifice, the IAEC should have achieved more and in a shorter time span if we would have managed to alleviate at an early stage the uncalled for taboo bureaucratic caps during the progress of the INNP.

The authors would have truly wished to pay tribute to each and every one of the Iraqi scientists, engineers, technicians and other personnel whose valuable contributions and endurance had shined throughout the duration of the INNP. However, they are too many to list. To all of them, Iraq is forever grateful, hopefully.

Finally, this testimony is intended to be homage and a tribute to the Iraqi National Nuclear Program that did demonstrate the ingenuity and persistent determination of Iraqi scientists, engineers and other supporting personnel; In addition to the to the adaptive capability of the program's managers in implementing such an ambitious and complex scientific program within the constrained environment of a developing country that was immersed in turmoil, once the decision was taken to go forward in implementing it, with the accompanying provision of the required funding and top level government support.

The authors wish to express their sincere gratitude to Dr. Abdel Qader Ahmad (Director of the Nuclear Research Institute and member of the 3000 Committee among other high level positions in the program) for his valuable edits and contributions especially to chapters one and two, and to Dr. Khalouk Refay (a lead physics research scientist in the Electromagnetic Isotope Separation (EMIS) method and later a general manager at the

Military Industrial Corporation) for his input and comments on the contents of this book as well as for his preparation of Annex 1.

1-1 Brief background on the Iraqi Nuclear Program

The Iraqi Atomic Energy Commission (IAEC) was established in 1956 in response to a U.S. gift, under President Eisenhower's Atoms for Peace program, of a small, yet relatively comprehensive, nuclear library and an experimental zero power reactor that failed to arrive due to the 1958 revolution in Iraq.

The gift of the nuclear library was, however, inadvertently potent and became a sort of catalyst to the Iraqi nuclear enrichment program three decades later.

The nuclear library was relatively extensive by the standards of that period. It included some of the Manhattan Project's atomic bomb research output. Starting around 1948-49 and during the early 1950s, a series of about fifty reports, books and opaque microcards (the precursors of the microfiche) were produced by the scientists and engineers who were directly involved in the American Manhattan Project that produced the first uranium and plutonium atomic bombs which annihilated two cities in Japan. These documents were published by the U.S. Atomic Energy Commission under the American National Nuclear Energy Series.

Only seventeen of these publications were reproduced as reports under the Technical Information Department, and were known as TID reports. The IAEC library had purchased a complete set of published TID reports on microfiche that included these 17 TID reports of the Manhattan Project in the nineteen seventies.

Moreover, eighteen of the National Nuclear Energy Series publications were published as hard cover books, each about 400 to 500 pages with detailed technical and scientific drawings and explanations. These books were part of the original 1956 library gift. They were later uncovered in 1987 stored in dust covered trunks in the IAEC library.

In addition to the above published materials, we found out in 1987 that the scientists and engineers who were involved in the Manhattan Project also filed 164 patents describing the details and drawings of the various processes related to their completed work on the Electromagnetic Isotope Separation method for uranium enrichment. These patents were publicly available, for a small fee, from the World Intellectual Property Organization in Geneva, Switzerland.

After the 1958 revolution in Iraq, the U.S. withheld its further support to Iraq in this field and the experimental reactor, which was approaching the Iraqi port of Basra in July 1958, was diverted to the nearby Iranian port of Bushehar and installed at Tehran University. In response, Iraq turned to the Soviet Union and contracted the purchase of a 2MW research reactor, labeled the 14th of Tammuz (July) reactor, with its supporting technological facilities. These were constructed in the Nuclear Research Center (NRC) at Al-Tuwaitha, 25 kilometers south of Baghdad. The reactor was operational in 1967 and was upgraded to 5MW in 1982.

The NRC became a magnet attracting fresh and seasoned Iraqi scientists and engineers who were mostly sent abroad, since the nineteen forties, to get their higher science and engineering degrees and who were either fully sponsored through government scholarships or were self-supported. Intensive plans were implemented to involve scientists, engineers and technicians in conferences, seminars, workshops and training all over the world as capacity building measures.

Research in the nascent NRC, with the collaboration of several International Atomic Energy Agency experts, was devoted to basic and applied research in nuclear physics, nuclear chemistry, biology, agriculture, and industrial and medical applications.

The first International Conference on the Peaceful Uses of Atomic Energy was hosted by the IAEC in Baghdad in April 1975. It was attended by a number of prominent scientists including the Nobel laureate, the late Pakistani scientist Prof. Abdus Salam, from the Abdus Salam International Centre for Theoretical Physics in Trieste, Italy who was the keynote speaker for the event.

1-2 Initial activities in reprocessing and other chemical lines of research

There is still a lingering and persistent question among those who are interested in the Iraqi Nuclear Program on the initial beginnings of Iraq's interest in uranium enrichment and whether or not it was initiated in 1981 with the subsequent acceleration of the scope and diversity of its research that culminated in the ambitious Iraqi National Nuclear Program (INNP) during 1981 - 1991.

It was, in fact, in the mid nineteen seventies, that the Director of the NRC, the late Dr. Khalid Saeed, had requested the Chemistry Department to start planning and implementing research in the following scientific endeavors:

- Chemical reprocessing (fission product chemistry and uranium-plutonium chemistry),
- Heavy water production,
- Stable isotope separation, and
- Helium chemistry.

1-2-1 Chemical reprocessing:

Though work on this line of research was requested, there was not yet sufficient scientific momentum at that time to carry out research on chemical reprocessing since the resources at the chemistry department at NRC, both human and technical, were nascent or not available. A number of fresh Ph.D. holders, who had just finished their degrees in fission and actinide chemistry embarked on literature surveys and performed some simple experiments on fission product chemistry in coordination with the Nuclear Research Center in Warsaw, Poland. Starting in 1977, and after a number of visits to the National Commission on Nuclear Energy (CNEN) in Italy, the first prototype Radio Chemical Laboratory was established in 1979 at the NRC. It contained a facility with the reprocessing capability of one burned fuel rod at a time, from the Russian 14th of Tammuz reactor. The process involved the dissolution in a hot cell of one fuel pin that was burned in the reactor core. The solution would then be transferred to another hot cell to separate the fission products. The remaining solution would again be transferred to another specialized glove box to separate the plutonium at a microgram level and finally the uranium using TriButyPhosphate (TBP) as extractant in the mixer-settler technique. This project was not publicly announced to the IAEC community. This process itself was supported by many lines of research at the chemistry department in the NRC.

Dr. Abdul Qader Ahmed, a leading chemistry researcher, was asked in 1976 to lead a team of reactor physicists to design and construct a sub-critical assembly using depleted uranium metal fuel rods for theoretical calculations. An attempt was made in 1977 to negotiate the purchase of 10 tons of depleted uranium fuel pins from Nukem, a West German company. Nukem turned down the request citing the constraint that its subcontractors in the U.S. and Canada had informed them that export licenses would not be issued for such material. Even though the plan was to construct a subcritical assembly as a research tool for the reactor physicists, it was wrongly construed to be used for other purposes.[9]

It is also noteworthy to mention that the IAEC had again requested Dr. Abdul Qader Ahmed, as Director of the NRC in 1984, to embark with new vigor on studying the requirements of a complete reprocessing cycle of the

---

[9] 1980 "Iraq places an order for 11,364 kg of depleted-uranium metal fuel pins from the West German company NUKEM. The pins are already fabricated into irradiation pins sized for the Osirak reactor and could be irradiated to yield plutonium. The 11 metric tons of target material are enough to produce 11kg of plutonium after 150 days of irradiation in the Osirak reactor. The deal is aborted when NUKEM subcontractors in the United States and Canada are told that export licenses would not be issued for the material.
—from one of the disinformation sources:"Iraq's Nuclear Weapons Program: From Aflaq to Tammuz," <http://nuketesting.enviroweb.org>; Frank Barnaby, *How Nuclear Weapons Spread* (Routledge, 1993), p. 91; Leonard S. Spector, *Nuclear Ambitions* (Boulder, CO: Westview Press, 1990), p. 187. http://www.nti.org/e_research/profiles/Iraq/Nuclear/2121_3291.html
Comment: The figures mentioned in the above abstract are strongly contested (the authors).

burned reactor fuel. The outcome of this effort by different departments of the NRC remained at the reporting level and was not implemented since most of the resources of the IAEC in the late nineteen eighties were devoted to the EMIS program.

1-2-2 Heavy water production

A simple investigation in 1981 was conducted into heavy water production starting with a visit to Romania to visit a prototype plant of about 1 Kg/day D_2O production rate. The method used was a counter current technique between water and hydrogen gas where the heavy hydrogen isotope (deuterium) from the hydrogen gas is to be exchanged with the light hydrogen in water resulting in increased concentration of the deuteron isotope in water. Heavy water is used as a coolant and moderator in reactors designed for natural, not enriched, uranium fuel. Visits were conducted locally to the oil refineries at Basra and Kirkuk searching for sources of hydrogen sulfide (H_2S) rich with deuterium isotope. However, all further efforts on this process were terminated due to the lack of proper facilities and equipment to measure deuterium levels in parts per million concentrations.

1-2-3 Stable isotope separation:

A series of experimental trials were attempted in the period between 1979 and 1981 involving the manufacture of simple ion sources at the Bader Establishment (a Military Industrial Corporation mechanical manufacturing facility near Baghdad). These ion sources were intended to be used in a miniature prototype spectrometer adopting a European Organization for Nuclear Research (CERN) design of a closed magnet for the separation of nitrogen isotopes in order to familiarize ourselves with this isotope separation process.

For this purpose, Dr. Salman Al-Lami, an electrical engineer, made several visits during 1979, 1980 and 1981 to CERN to be acquainted with the magnetic field calculations of this separation process. He was poisoned by a hitherto mysterious virus, assumed to have been delivered by Israeli Mossad agents. Despite the Iraqi Atomic Energy Commission sending a medical team to Geneva to halt his disfiguring symptoms, Dr. Salman Al-Lami died in Geneva in 1981. He was the first victim of the Israeli Mossad which has assassinated several other members of the Iraqi nuclear team.

Originally, it was thought to use UF_6 as the feed material for a future prototype isotope separator and to start with slightly enriched uranium. For this purpose, a highly secret mission from the IAEC was delegated to China in November 1980 to negotiate with the relevant Chinese authorities for the supply of five tons of UF_6 with 5% enriched uranium. During a week of tough negotiations in Beijing, all details, including transportation and airway route from China to Iraq were discussed and agreed upon. However, the Chinese failed to deliver the consignment using the commencement of the Iran-Iraq war, which erupted in 1980, as a pretext. In any event, it was later decided to use UCl_4 instead of UF_6 as the feed material to the isotope separator magnet.

1-2-4 Helium chemistry

Work was abandoned on this line of research for lack of relevant facilities.

1-3 A scientific and technological leap forward

In 1976, the late president Saddam Hussain presided over the purchase of two experimental French Osiris class reactors to also be located at Al-Tuwaitha. The reactor Tammuz 1 was a 40MW material testing light water experimental reactor using 93% enriched uranium fuel, while Tammuz 2 was a 0.5MW mockup of Tammuz 1 for mapping the distribution of neutron fluxes in the reactor core and for testing the experimental rigs before fully employing them in Tammuz 1. The full scope of this project was under the complete surveillance of the International Atomic Energy Agency.

Three years after the signing of the contract, France suggested to Iraq that it recommends fueling the reactor with the newly developed low uranium enriched fuel type called "caramel". The then deputy chairman of the IAEC commission, Dr. Abdul Razaq Al-Hashimi, refused to accept this proposal calling it a breach of the contract terms. Eventually, only 12 kg of the highly enriched contractual fuel was delivered to Iraq.

On June 13th 1980, Dr. Yehia Al-Meshad, an eminent Egyptian nuclear scientist who was working for the IAEC, was assassinated in a Paris hotel room while on a mission to France. The Israeli Mossad was again highly suspected of carrying out the assassination.

Work commenced smoothly on the construction of the two research reactors at Al-Tuwaitha and the training of approximately 60 Iraqi personnel at the Saclay Nuclear Research Center in France, despite several sabotage attempts of blowing up key reactor components in a French port by Mossad agents and their assassination of a number of scientists and engineers. The Mossad had undoubtedly attempted to make other contacts with some of the about 60 Iraqi scientists, engineers and technicians who were undergoing training at Saclay. Tzipi Livni, who recently revealed herself to have been a Mossad agent, was stationed in Paris during 1980-1982 and who is at present the leader of the Israeli Kadima party, did try to establish contact and foster a relationship with Dr. Imad Khadduri under the guise of conducting a consumer survey with Dr. Imad and his wife. However, her attempt was aborted, probably upon being informed over a dinner invitation by her of the previous experience of her potential recruit in armed struggle against Israel. It is not known who else was approached among the Iraqi contingent in France at the time. However, it is certain that the Mossad had intensified its efforts to abort any progress in Iraq's nuclear endeavors despite international surveillance and agreements.

The initial loading of the nuclear fuel and the startup of the reactors was destined throughout the summer of 1981, and the startup of the reactors was intended during the fall of 1981. The French highly enriched fuel consignment that arrived in Iraq in early 1981 was registered legally with the IAEA according to the requirements of the Non-Proliferation Treaty (NPT) by Dr. Abdul Qader Ahmed who went to Vienna for that specific purpose along with the Scientific Attaché to the Iraqi embassy, Suror Mirza Mahmoud.

On Sunday the 7th of June 1981 at dusk, the Israeli Air Force launched a surprise strike on Al-Tuwaitha Nuclear Center with eight F-16 multi-role jet fighters and six F-15s for escorts. The direct strikes severely destroyed the Tammuz 1 French reactor and to a certain extent the Tammuz 2 reactor. Notably, the smaller reactor was at the time of the strike fully loaded with nuclear fuel. A few of the bombs came marginally short of hitting the core of the reactor, which indicates clear intent to strike the core. Had the directed bombs actually fall on the core of this reactor, there would have been a serious radiation contamination to the surrounding area and Baghdad.

During the attack, Dhafir Selbi was in close proximity to the bombed area and arrived at the site within 15 minutes of the bombardment. The late president, Saddam Hussain, had sent the Defense Minister at the time, the late General Adnan Khairallah, that same evening for a firsthand look at the damage and to carry out an investigation and to report his findings. However, it was not possible to make a close inspection of the damaged area since the surrounding grounds were littered with Israeli cluster bombs. That was a further clear evidence that Israel was not only targeting the project itself, as they had publicized, but they were intent on causing the maximum possible damage, including targeting the Iraqi personnel who would have been present at the site or inspecting the damaged site later on.

The military cadre, under the direction of General Adnan Khairallah, commenced that same night on gathering the scattered cluster bombs in piles, covered them with sand bags and detonated them. General Adnan Khairallah stayed for a few hours during which he started to organize the launch of the required investigation into the details of the attack.

Dr. Abdul Razaq Al-Hashimi, the deputy chairman of the IAEC at the time, and Dhafir Selbi attended part of the military investigation in an unofficial capacity. During these investigations, it transpired that the intrusion of the Israeli airplanes was, in fact, detected by the Iraqi air defenses in the west of Iraq. However, the conveyance of that message from the central air defense command in Baghdad to the south of

Baghdad sector of Al-Tuwaitha was inadvertently delayed. Had it been acted upon in time, it would have been almost certain that the Israeli fighter planes would have suffered heavy retaliation before reaching Al-Tuwaitha and would have probably not reached their target.

After the conclusion of the investigations, the late president met with the commissioners of the IAEC to discuss the aftermath of the strike. Dhafir Selbi was present at that meeting and he recalls that the strike had developed a relentless posture in the late president's conviction that he would seek revenge for the incident, sooner or later. A reason why this particular information regarding the delayed warning message had remained classified, until now, was in order to keep Israel smug in the assurance of its capability to carry out future air strike attacks against Iraq using a similar maneuver that would be foiled as a result of its arrogance.

Chapter 2: The Iraqi National Nuclear Program

2-1 The start of the Iraqi National Nuclear Program

Provoked by the Israeli aggression on the Tammuz reactors, Dr. Humam Abdul Khaliq, the deputy chairman of the IAEC commission at that time, asked a team of several NRC scientists and engineers in the summer of 1981 to make a concentrated literature survey on the possibility of embarking on a national nuclear program for the enrichment of uranium.

The task of this small group of scientists and engineers was to come up with the requirements of the resources and capabilities in order to undertake such a program and its anticipated duration. The team was given a time frame of a few weeks to submit its report. The team relied mainly on the literature that was available at the IAEC library and gleaned the publicly available information on the required research paths, from a scientific and engineering point of view, and handed in the report within the given time deadline.

During the same period, several books and reports that were related to the subject matter were sent to Dr. Jafar Dhia Jafar during his house arrest for his consideration. Dr. Jafar was tasked/volunteered with the preparation of his own report on the requirements for a national nuclear program on uranium enrichment which was to be executed solely by Iraqi scientists and engineers. Within a few weeks, Dr. Jafar had compiled his report that contained theoretical calculations on the separation of the uranium isotopes using different methods and the possible enrichment yields. The report was submitted directly to the late president who sent it in turn for consideration to the then deputy chairman of the IAEC, Dr. Abdul Razaq Al-Hashimi, who in turn asked Dhafir Selbi to join him in the deliberations of its obscure theoretical calculations and the anticipated enriched uranium yield implications. Dr. Jafar was shortly thereafter released from his house arrest in early September 1981 and he recommenced his work at the Nuclear Research Center. It might also be worth mentioning that Professor Abdul Salam intervened in the release of Dr. Jafar by preparing a letter during that period that was passed along through Dr. Abdul Qader Ahmed (the Director of the NRC from 1982-1986), requesting higher authorities to release Dr. Jafar. This may have also contributed for his eventual release.

Dr. Jafar resumed his position at the IAEC and was soon appointed the head of the newly founded Directorate 3000. He immediately gathered a group of selected scientists and engineers and asked them to repeat their literature survey effort by focusing on the available 17 TID reports on the EMIS enrichment process and to further investigate various other methods of enrichment of uranium that were known at that time.

In the meantime, a security guideline was put in place to limit the access to information with a circular boundary of access, where the few people in the inner circle had full access to the gathered information and the outer circles had diminishing access to it. The nominal slogan for the plan was "information is provided on need to know bases".

2-2 Attempted uranium enrichment technologies

The results of the concerted effort by Dr. Jafar and his selected team of scientists and engineers in determining the viable research paths on uranium enrichment resulted in recommending the adoption of the following uranium enrichment methods:

2-2-1 The gaseous diffusion method

This method entailed the use of a special barrier of porous material through which uranium gas would be forced to penetrate separating the slower moving heavy uranium 238 atoms from the nimbler and lighter uranium 235 atoms.

Starting in 1982, attempts were made to manufacture such a demanding barrier but these efforts were met with very little success. The process also required compressors capable of working with the highly corrosive uranium hexafluoride gas (UF$_6$) being the process media at higher than room temperature. Despite attempts made to manufacture such a compressor using reverse engineering (after purchasing a number of such compressors from the USA under covert guises), the efforts resulted in little success. This was especially detrimental as cascaded units were essential to acquire a substantial amount of enriched uranium and their purchase would be indicative and difficult.

Furthermore, this method faced yet another unattainable technological obstacle as it required the development of sophisticated pyro-electrolytic cells for the production of fluorine gas (F$_2$) which was needed for the preparation of uranium tetraflouride (UF$_4$) and uranium hexafluoride (UF$_6$). Both of these gases required equipment and materials that can survive the severe corrosive conditions for their preparation and production which were beyond the capabilities of the IAEC and Iraq in providing at that time. This method was abandoned after a few years.

2-2-2 The gas centrifuge method

The Main effort on this method of uranium enrichment commenced belatedly in 1987 under the umbrella of Hussain Kamel, the Head of the Military Industrial

Corporation, with a group that was separated from the IAEC, as will be further explained in Chapter 3.

Despite the stated directives in the summer of 1981 for the need to pursue uranium enrichment solely with Iraqi nuclear expertise and personnel in the belief that the risk of our exposure in this effort, if we would have involved foreign assistance, would far outweighed the likely technical benefits of that foreign assistance; yet this policy was not adhered to by the group that was separated from the IAEC and transferred to Hussain Kamel who were tasked with the gas centrifuge enrichment process. This group purchased many designs, drawings and specifications related to the required centrifuge machines from a German firm. Some progress was achieved in reaching fast spinning speeds of the prototype centrifuge machine.

The work on this method continued without fruitful results till the end of 1990. The German drawings were deliberately hidden without the knowledge of the Iraqi authorities for twelve years. They were only produced again by the lead person of the group, Mahdi Obeidi, who hid them in his garden, until after the occupation of Baghdad in April, 2003 and delivered them to the occupation forces in order to secure his relocation and his family to the U.S.. These same drawings were one of the last three remaining 'outstanding' items to be delivered by Iraq, despite the repeated calls by the Iraqi authorities for the INNP community to hand in all documents to the IAEA inspectors, that were clung to by the IAEA inspectors as an excuse for their reluctance to announce the completion of their inspections and to lift the UN economic sanctions on Iraq regarding the transparency of its earlier weapons of mass destruction programs. Although those three outstanding items were in fact only mere marginal fragments of the devastated program, yet the IAEA leveraged them as a pretext for asking for an additional period of six months of inspections.

It is believed that the American occupation force relocated Mahdi Obeidi and his family to the U.S. in the hope of finding other documentation on weapons of mass destruction that would justify its decision to invade Iraq and not these engineering drawings per se which they could have easily obtained from the manufacturing company in Germany and in any event would represent a marginal need for Iraq to re-embark on a project that uses this technology. This was indicative of the desperate measures to which the American administration went in its effort to manifest any proof that might justify its false claims that Iraq had weapons of mass destruction as a justification for its occupation. Hence, Mahdi Obeidi did exploit, for his own personal benefit, the desperation of the occupier to produce any evidence to justify its occupation. On the other hand, it is to be noted that the Americans, with all of their intelligence and scientific resources, wouldn't publicly come to the conclusion that what Obaidi had would be useless in their quest to pin Iraq on the WMD issue.

2-2-3 The laser isotope separation method

The laser group in the Physics Department of the NRC received a directive from deputy chairman of the IAEC in the fall of 1981 to commence work on Laser Isotope Separation as part of that department's scientific endeavors. Research in this uranium enrichment method involved both molecular and atomic vapor technologies. It also entailed a number of other activities relating to the manufacturing of laser components for use in laser-related experimentation, particularly the CO_2 lasers.

However, when the achievements of the laser group were appraised in 1987 it became apparent that this line of research had not reached a satisfactorily point of an integrated experiment that would be able to achieve isotopic separation of either elemental uranium or UF_6. This was also affected partly by the fact that the process was given neither higher priority nor resources due to the deep involvement with the EMIS program. In addition, there also appeared specific evidence that threw in doubt the accuracy of the originally claimed data and information that were relied upon to initiate this activity. Finally, this line of work was hence terminated in 1988.

2-2-4 The chemical uranium enrichment program:

This program started in late 1987. It focused on investigating both the ion-exchange process that was being developed by the Japanese and the liquid-liquid solvent extraction process that was being developed by the French.

A unique combination of both the ion-exchange and the solvent extraction techniques was successfully realized on an experimental level to achieve potential enrichment levels of about 11%. This could then be used in the preparation of the initial feed material for the electromagnetic separation process and to increase the quantity and quality of the final product. It would also serve to decrease the number of required stages needed to reach higher uranium enrichments.

One of the main obstacles for a faster pace of development of this enrichment process was again the limited availability of engineering and manufacturing support for their research and development requirements since top priority, by the end of the nineteen eighties, was

36

given to the electromagnetic separation method as will be explained in the coming section and following chapters.

It is also worth mentioning that in 1980, as part of their proposed deal to switch the Tammuz reactors' fuel to the low enriched caramel, the French had alternatively proposed the sale, for US$50 million, of the detailed design of their pulsed column based Chemex process to Iraq. The process would allow for the production of low uranium enrichment fuel for nuclear reactors. This offer did not mature to a deal status due to the fact that Iraq had insisted on and obtained some of the highly enriched uranium fuel that was contracted for in the reactor supply deal.

In addition to the above chemical extraction activities, a novel promising line of work was developed by using crown ether as an extractant for the separation of uranium isotopes. This method, however, was confined to a laboratory scale only.

2-2-5 The Electromagnetic Isotope Separation (EMIS) method

As mentioned earlier, preliminary research on this method had started even before 1981 in an overt modality and aim. It entailed the prototype manufacturing of an ion source in a closed type magnet for the stable isotope separation method in the facilities of CERN (The European Research Center in Geneva).

The electromagnetic isotope separation method is based on the principle that uranium ions of the same energy, but of different masses, describe trajectories with different curvatures in a magnetic field. In this process, uranium atoms are ionized, extracted then deflected in a magnetic field. The two deflected separated ion beams of interest (U_{235} and U_{238}) are then collected in two different collectors.

Work on the EMIS process was focused upon right from the start of the INNP in 1981 due to its simplicity, ease of manufacturing of the magnets and the already available scientific literature from the Manhattan Project. However, the choice of the type of the ion source presented the greatest obstacle in employing this method on a production scale. This issue, with its accompanying technical obstacles, as will be explained later in Section 4-1-2, led to the consuming of essential time and resources until it was resolved decisively in 1987.

The need for the choice of the final type of the ion source spearheaded the complete administrative overhaul of the management of the INNP and catapulted its progress into high gear. It was during that transition phase that the INNP acquired new objectives, including the commencement of work on the design of the "device"; i.e. the atomic bomb mechanism without its nuclear charge.

2-3 Uranium enrichment using the EMIS method

The enrichment method that was chosen for the production state was the EMIS method that was employed for the Manhattan Project. The main novel part of the INNP contribution to the EMIS process was the development of a double focused field magnet and the development of state of the art control systems, in addition to several other contributions that will be covered in the next chapters.

Some of the main scientific INNP achievements in this technique were:

- The precise measurement of the magnetic field by a novel electro-mechanical engineering apparatus that was employed to map the magnetic field in the angular and transverse directions using Hall-effect sensors placed in horizontal and vertical positions on a moving arm.
- The formation and control of plasma beams. The first uranium separation occurred in January 1986. While the first measured uranium current was 1mA with an availability factor of 10-20% from the separators at Al-Tuwaitha, it reached a current of more than 120 mA in

1990 with an availability factor of 46% from the larger 120 cm mean radius units installed at Al-Tarmiya.

See Annex 1 for a technical summary of the various stages in the development effort of the EMIS program in the INNP.

See Annex 2 on the engineering infrastructure that was created to support the INNP.

Chapter 3: Two Important Milestones in 1985 and 1987

3-1 An unfounded promise in April 1985

While the Iran-Iraq war was raging in April 1985, the members of the executive committee (i.e. the Commissioners) of the Iraqi Atomic Energy Commission (IAEC) at the time: (Dr. Humam Abdul Khaliq (also deputy chairman of the IAEC), the late Dr. Rahim Al-Kittal, , the late Dr. Khalid Saeed, Dr. Muyasser Al-Mallah, Dr. Jafar Dhia Jafar and Dhafir Selbi (the latter two were appointed during 1982) were notified that they were to attend a meeting with the late president Saddam Hussain and that they will be informed in due time of its time and venue.

The agenda for the meeting was to review the status of the ongoing projects in the INNP.

On a following evening, the commissioners were requested to be present early next morning at a meeting point where they would be escorted to the meeting with the late president.

Once assembled at the assigned meeting point, the convoy of cars, led by the ex-vice president Izzat Ibrahim (who was also the Chairman of the IAEC at that time succeeding Saddam Hussain once he had become president of Iraq), headed towards Al-

Radwanya near Baghdad's airport where several of Saddam Hussain's palaces were located.

From the security checkpoint leading to the Al-Radwanya complex, they were led to a large mobile caravan. Inside it, they found the late president seated at the head of a meeting table with his personal secretary, Hamed Hummadi, seated next to him. They shook hands and were seated. The late president started the meeting enquiring how everything was going with the commissioners, which is the customary polite opening dialogue. At that point in time, Hussain Kamel, who was then the head of the security teams guarding the late president, did not attend the meeting and had stood at some distance from the caravan with the security entourage.

Following the pleasantries, the IAEC deputy chairman started presenting the status of the projects that were ongoing in the INNP. The report that he was reading from was jointly prepared by Dr. Jafar Dhia Jafar and himself. It had not been previously circulated to the other attending IAEC commissioners.

After a lengthy exposé of the current activities at that time, the IAEC deputy chairman concluded the presentation with the surprising and potentially binding announcement that the INNP will be achieving its "fruitful objectives" in 1990; i.e. in five years time hence. It was not made clear, at least to the other attendees, exactly what the objectives were, to which he was referring. Implicitly, the other commissioners inferred that the promise could have alluded to Iraq's attaining of a new milestone in its INNP as a whole and its nuclear enrichment capabilities, taking into consideration the moment in time the meeting was being held and the circumstances under which it was held.

The suddenness of the announcement electrified the atmosphere of the meeting. Most of the attendees were taken by surprise. Once the gravity of the announcement sank in, the late president's eyes had visibly swelled with emotional tears.

The late president then regained his calm, turned to the attendees and enquired if any of them had any comment on what was presented by the report. Since the announcement was a total surprise for the listening commissioners, except for Dr. Jafar and the IAEC deputy chairman, there were no comments made.

This was indeed a tall order of a commitment since most of the EMIS activities that had been attempted during the preceding four years, in as much as the design and the manufacture of the constituents of the EMIS process and thence the mastering of the process parameters had still not progressed more than the laboratory experimental scale stage (as a matter of fact, actual separation was first achieved in 1986, as mentioned earlier). Neither had any industrial scale prototype project been launched.

The impact of this promise must have been profound on the late president as Iraq was engaged in a vicious war with neither an end in sight nor any clear indication of who the winner would be. It was also during the time of the "Iran Contra" affair that it was indicated, if anything, that the U.S. was intent on prolonging this war without a clear victory in sight for either side.

Considering these overburdening circumstances, the news of the promise appeared to have been immensely appreciated by the late president in that he could not stop the tears from rolling down from his eyes.

A decision was to be taken based upon existing premises and their extension into the near future of that time. Hence, the invasion of Kuwait in August 1991, when the promise was to have been fulfilled, must have been juxtaposed with the fact that nobody dared to tell the late president that the promise that was made earlier was not going to be fulfilled. Had he been told, this may very well have prevented him from taking the decision to invade Kuwait as it might have enabled him by then to possess an unconventional deterrent weapon that would force any opponent to the negotiating table instead of confrontation. This interpretation might also explain the hurried effort to extract enriched uranium from the available nuclear fuel after the invasion of Kuwait in 1990, as will be detailed in section 5.5.

It is abundantly clear that since the invasion of Kuwait in 1990, the situation in Iraq began to drastically deteriorate till the occupation of Iraq in 2003 that was followed, and until now, by widespread terrorism, sectarian violence and a rudderless future. This might not have happened had Iraq not invaded Kuwait. This would be a lesson for future generations.

The meeting with the late president ended shortly thereafter. There was a pad in front of each commissioner to jot down his comments during the meeting. These pads were collected by Hamed Hummadi at the end of the meeting upon which he wrote on each pad the name of the corresponding attendee.

One would wonder if this promise had any link with promises attached to the report that was prepared by Dr. Jafar several years earlier while he was under house arrest. Another important outcome to this grave and sudden announcement is what transpired two years later when the promise was declared unachievable, as will be explained in the next section.

It is important to emphasize the repercussions of this promise on the late president at that difficult juncture, as the war with Iran was at its peak without any foreseeable resolution. Added to that impasse was the depletion of Iraqi resources that had forced Iraq to be heavily indebted and to borrow heavily coupled with the desperate attempt to not engage any outside world power to intervene to resolve the conflict On the contrary, a conviction was arrived at that some of the world powers were not concerned with the dragging conflict any more while some powers were in fact covertly and overtly providing resources for its prolongation.

All of these factors must have played a major role in the thinking and planning of the war in the mind of the late president to the degree reflected by the tears in his eyes. We do not think that those who made the promise had considered its impact before announcing it. Had they considered such a dimension to the promise, they might have at least waited till we actually achieved successful enrichment results which only occurred on a modest laboratory scale about a year after the promise was made. It is probable that the enormity of the promise and its ramifications did finally sink in which might explain the reluctance to inform the late president of its failing as will be shortly explained.

After that meeting, the late president bestowed upon each of the commissioners, as well as to other Director Generals in the IAEC, a new Mercedes car and a mobile caravan as gifts. That startling generous gesture started all sorts of rumors amongst the IAEC employees who wondered why were such gifts suddenly lavished as there was no indication on the ground of any great breakthroughs in any of the ongoing programs.

Back in the office of IAEC deputy chairman at Al-Tuwaitha site in the afternoon of the same day of the meeting with the late president, the late Dr. Rahim Al-Kittal, in the presence of Dhafir Selbi, confronted the IAEC deputy chairman regarding the event of that morning. He pointedly questioned the credibility of the suddenly announced 1990 target date. He demanded to know how the date was arrived at and what would have been the basis for it. He also posited the gravity of not pre-involving the rest of the commissioners in such an important commitment beforehand while their mere presence in that meeting with the late president did hold all of the commissioners responsible in the anticipated delivery on such an important promise.

The ensuing discussion quickly became heated accompanied by raised voices so much so that the IAEC deputy chairman felt it necessary to leave the office in order to ask his secretary to make sure to clear the surrounding corridors lest passersby would hear the raised level of shouting and the content of the discussion. Both of them became very agitated. They had to end the meeting abruptly on a sour note in order to avoid further stressful confrontation.

That morning's encounter was the first meeting ever in which a clearly expressed commitment was offered by the IAEC to the late president on a possible timeline for the INNP. However, the exact nature of the promised outcome was still not clarified to the other commissioners, though it was a wide brush promise implicitly entertained by them.

3-2 Ramifications of the failed promise - two years later

During the period of 1981-1987, the involvement of the IAEC commissioners in the Iraqi Nuclear Program continued to be relegated to the routine hearing of the progress report presented to them at the end of every quarter of the year by the

IAEC deputy chairman and by Dr. Jafar, head of Directorate 3000, which concerned the general features of the various enrichment projects that were being covertly implemented in the INNP. The rest of the commissioners were scantly involved in the details of the program itself and were usually informed of the progress of the various ongoing projects in general outlines that were not coupled to target values or measured outcomes nor to scheduled dates. These summary presentations were capped by equally general comments from the rest of the commissioners that would only relate to the broad issues that emerged from the sketchy presentation.

In one such quarterly meetings, in April 1987, and exactly two years to the month after that fateful promise by the IAEC deputy chairman and Dr. Jafar to the late president, Dr. Jafar was presenting the Directorate 3000 progress report for the first quarter of 1987.

The presentation at that meeting was unexpectedly detailed. Dr. Jafar was nervous and he repeatedly invoked the IAEC deputy chairman in the deliberations of the progress of the programs. It did appear that the IAEC deputy chairman was already briefed about the report by Dr. Jafar and was apparently dissatisfied with its somber conclusion.

At the end of his presentation, **Dr. Jafar announced that the promise made to the late president earlier in April 1985 could not be kept**. The IAEC commissioners were shocked by this sudden and unexpected declaration bearing in mind that all main IAEC resources were already committed and allocated to the EMIS program at the expense of other related research options that were accordingly put on an idling status.

The impeding reasons that he offered were mainly related to the lack of obtaining reproducible and reliable experimental results due to the lack of knowledge and expertise, and hence the mastering of some of the essential constituents of the technology.

The first commissioner to comment on this revelation with an offensive tone was the late Dr. Al-Kittal. He explicitly pointed out that such a backtracking, which he had always warned of, and especially during the meeting with the IAEC deputy chairman after the meeting with the late president in April 1985, had indeed gone unheeded. He also wondered why the just presented conclusion by Dr. Jafar was not conveyed to the late president in the same manner in which the April 1985 promise was made to him then.

That was the most stressful and nerve wrecking meeting that was attended by the IAEC commissioners.

The heated meeting itself dragged on for hours. The commissioners all realized that they must find a solution to this impasse since no one dared to convey such a failure to the late president after the lapse of 40% of the time that was promised to meet the target during the 1985 April meeting. Everyone who was attending that meeting shuddered at predicting the late president's reaction to this state of affairs.

Furthermore, and according to the knowledge of one of the attendees, namely Dhafir, this back tracking was never officially conveyed to the late president in so many words by the IAEC that would have clearly indicated this backtracking.

As the heated discussions continued during that fateful meeting, the commissioners requested Dr. Jafar to pin point for them the nature of his difficulties. They wanted to help but only if they knew the exact nature of the difficulties that he was facing. He finally posited that he was very distracted by the heavy administrative work load expected from him at the expense of concentrating his attention on the core technological aspects of the projects in order for them to move forward.

As the stressful meeting dragged on for hours, the commissioners were becoming very exhausted to the point that they were not sure if continuing the discussions could have been of any further use.

At one of the junctures at the end of that meeting, Dr. Jafar repeated in more detail the drag effects of his administrative load and its hindrance in allowing him to give enough attention to his core scientific responsibilities.

At that point, Dhafir volunteered to assist Dr. Jafar in handling the administrative affairs of Directorate 3000 for a few months in addition to his heading of Directorate 4000 which was at that time responsible for the general administrative chores of the other major departments at the IAEC as well as the department of Engineering Services.

One could hear a big sigh of relief at this proposed solution and the recommendation was immediately approved unanimously by all attending commissioners.

At that point in time Dhafir did not know in any sufficient detail the full picture of the scientific and engineering advances, or lack thereof, that were going on in the INNP. Nonetheless, he was convinced that if he gave Dr. Jafar the administrative leverage that he needed, it would relieve him to accelerate the enrichment programs. However, and upon assuming the new responsibility, Dhafir found himself getting more and more involved into the scientific and engineering policies and decisions of EMIS ever more deeply as time progressed.

In cognizance of this new role, ex-vice president Izzat Ibrahim issued a long letter of thanks to Dhafir and to Dr. Jafar in April 1987 which was read during a gathering for the leading cadre of the IAEC by deputy chairman at the IAEC rest house overlooking the Tigris river at Al-Tuwaitha.

Later events indicated that the ex-vice president Izzat Ibrahim (chairman of the IAEC) was himself not told of the real reasons for Dhafir's volunteering to assist Dr. Jafar. He was merely informed that the shift of responsibilities was implemented in order to give the highest priorities to the scientific activities of the INNP and in rallying all possible available efforts and resources.

Chapter 4: Organizational Milestones

4-1 The lifting of taboos

In addition to the April 1987 milestone meeting mentioned above, the years 1987-1988 witnessed other crucial events in the short life of the INNP that resulted in unprecedented surges and leaps in its level of activities, including advancement in the prototypes R50 and R100 separator units.

In the aftermath of the April 1987 heated commissioners' meeting and after preliminary investigation of how the configuration of the administrative hierarchy was functioning and how the channels of communications were being conducted in Directorate 3000, Dhafir suggested that the leading cadres in Directorate 3000 would hold a retreat at the Al-Habanya resort near Fallujah for a few days to evaluate the state of affairs in Directorate 3000. There was an urgent need to specifically consider whether the prevailing organizational structure was an optimal configuration that can efficiently process the diversified activities of the Department in an interactive and effective manner and in facilitating the advancement and coordination of the projects to the maximum possible extent. More importantly, there was a need to pool together and utilize most efficiently the available human resources and the material resources that were made abundantly available in financial terms to the INNP.

Dhafir had always maintained that an efficiently designed administrative organization would facilitate the maximization of the use of the potential of all the participants' inputs, by smoothing the flow ability of their collective product. This might sound obvious, but in reality it does require a formidable effort to put together such a structured organization as it does require deep insight of many diversified aspects of management and cultural sensitivities.

Furthermore, since the quantity and quality of the scientists and engineers that was available to the INNP was in fact at the short end of the scale of what should have been made available for such demanding complex projects that need to be completed in short time frames as compared, for example, to the resources and talents that were made available to those involved in the Manhattan Project. These included twelve Nobel laureate scientists in physics and chemistry, such as Albert Einstein[10], Enrico Fermi, Ernest Lawrence and Arthur Compton, in addition to the engagement of well established industries with extensive experience in design, manufacturing and operation, such as Kodak and Ferguson. Hence, the maximization of all participants' output should always be the top priority of the management.

Dr. Jafar and Dhafir headed the introspective meetings that were held for a few days in Al-Habanya resort during May 1987.

---

[10] Einstein was not directly involved in the Manhattan Project. He did alert President Roosevelt on the power of the atomic bomb at the start of the World War II and he issued, jointly with Bertrand Russell, the Russell-Einstein Manifesto on the danger of the use of the atomic weapon.

The atmosphere during that retreat was very conducive as the weather was pleasant and the discussions candid. Families of the participants were also invited which added a social dimension to the gathering. Furthermore, the security of the gathering was reasonably arranged and without undue fuss.

The scientists and engineers who participated in this retreat found in it an opportunity and a chance for a fruitful openness. They openly aired their smoldering grievances. It also relieved, to a large extent, their mounting frustration of being required to tackle very demanding tasks with limited human resources. Their dissatisfaction was compounded by the burden of a sluggish organizational set up which led to endless meetings whose outcomes did not result in clear and positive results and usually ended with exchanges of bitter recriminations.

After that sojourn in Al-Habanya, the participants went back to work at Al-Tuwaitha with rejuvenated spirits. Dhafir started holding one-on-one meetings with several of the leading personnel in Directorate 3000, starting with Dr. Jafar. The aim was to gather further details that might assist in the reconfiguration of Directorate 3000. After two weeks of interviews, he arrived at a clearer understanding of the existing work flows, the organizational draw backs and the changes needed for an optimized organizational reconfiguration.

It had become apparent that Directorate 3000 had suffered over the past several years from malaises that included failed approaches, mismanagement, unproductive organizational structure and unconsolidated plans, the cumulative results of which were manifested in the following symptoms:

4-1-1 Ineffective approach and procedures in the design process

A main flaw in the administering of the research projects until 1987 was that it had emulated previous management practices that lacked timely interactivity between the different disciplines; primarily between the physicists and chemists (the up streamers in designs and activities) and the different engineering disciplines (electrical, mechanical, electronics, etc..) who were tasked with the drawing up and the issuing of the designs of the enrichment processes' constituents and systems. The required design had usually commenced with the basic requirements by the physicists and chemists. It then used to flow in only one direction, from one engineering discipline to the other. At every stage, each of the engineering disciplines would **separately** input its own thoughts and comments on the proposed design requirements that mostly ended with Dr. Jafar for a decision to implement, until it finally arrived to the manufacturing entity where it was to be manufactured. The resulting manufactured constituents/systems would then be handed over to the process people to implement and operate.

However, upon actual implementation of the manufactured products in the process, the operators most often did face unexpected results that would barely match the original objectives that were laid out by the physicists and the chemists.

Such often repeated failures had given rise to all sorts of heated and time wasting recriminations amongst the concerned parties that had participated in the formulation of the design on the one hand, and the manufacturing and fabrication departments on the other hand. The personal acrimonious discussions would usually lead to no avail. The heated discussions had also heightened the frustrations of the participants who had to start all over again, with reduced vigor, to reformulate the required design from scratch. Such frustrations had taken a heavy toll on most participants to the degree that most leading personnel were hopeless of ever coming up with any viable solutions.

4-1-2 We should have started where others ended

A hindering outcome of the above divergent approaches solidified on the attempted realization of a properly working Penning Ionization Gauge (PIG) ion source which was at the heart of the EMIS since 1981. Its employment was favored at the start of the INNP due to the fact that the PIG ion source promised to be an advanced ion source version for the EMIS, in comparison with the calutron ion source that was employed in the Manhattan Project, since the PIG employed an anti-cathode to increase the degree of ionization, and it utilized an indirectly heated cathode to protect the hot filament from the corrosive UCl_4 vapor.

The bifurcating positions concerned the fact that the PIG ion source had manifested marked instability in its operation due to the fact that short-circuits had often occurred between the anode and the anti-cathode primarily due to its complex configuration. These repeated failures resulted in the non-reproducibility of the ion yield. This was amplified by the fact that no theoretical calculations were performed to understand the low-pressure discharges and plasma behaviors in the presence of magnetic fields and, hence, could have guided progress in the design of the PIG.

Unfortunately for this complex system, an experimental approach was solely and persistently employed over the years by several leading scientists. This approach was particularly employed in the definition of some ion source functional parameters, in order to improve its performance, and contrary to a sound scientific approach that would have used advanced available technologies in order to shed light on theoretical premises that would facilitate insight and action for the way forward .

The persistent technical difficulties led several scientists and senior engineers to repeatedly recommend the use of the relatively less complicated calutron ion source that had already been employed successfully on an industrial scale level in the Manhattan Project. The calutron's previously recorded performance had exhibited more reliability and promising yields. Their requests to adopt the calutron instead of the PIG ion source were repeatedly brushed aside.

The repeated disregard for the adoption of this alternate option relegated it to pointed jokes. The elevation of the principle of continued trial and error in the design of the ion source to a sacrosanct level, at the expense of experiments based on theoretical simulations and calculations[11], became the main subject of puns in side chats. The lingering complaint was why should we flog to death the PIG ion source with endless trial and error attempts and developmental experimental approach to its design with no theoretical attempts to mathematically model the ionization, while the TID reports of the Manhattan Project referred to proven experimental designs of the calutron which should have been adopted and further developed?

Therefore, the issue of the switch of the ion source from PIG to the calutron had become a focal point of the required change, primarily due to the failure of the PIG over years of experimentation to provide reproducible results.

In a meeting in early 1987 that was attended by Dhafir Selbi and senior engineers the late Basel Al-Qaisi and Dr. Qais Numan, it

_____

[11] Trial and error was perhaps the only viable conduct during the Manhattan project in the nineteen forties with tight war time constraints and without the availability of computers. However, to adhere strictly and solely to trial and error experimentation in Iraq during the early years of the INNP in the nineteen eighties was not justified with the availability of computerized theoretical modeling possibilities that offered relevant potential insights. This venue should have been given higher priority; yet, it was not seriously indulged in. As will be elaborated in the next section, we had a potentially important resource of a very qualified theoretical physicist who could have developed relevant software which would have offered valuable theoretical insights into the ion source design from the early stages of the INNP.

was pressed upon Dr. Humam to intervene in resolving this stalemated matter and to allow for the use of the calutron ion source. This step was finally decided upon in a larger meeting that was attended by all of the concerned heads of activities in Directorate 3000; namely, Dr. Humam, Dr. Jafar, Dhafir Selbi, Dr. Khalouk Refay, Dr. Thamer Numan, the late Basil Al-Qaisi (Head of electrical and electronics engineering department), Zuhair Al-Chalabi (Deputy Head of electrical and electronics engineering department), Munqith Al-Qaisi (Head of the mechanical engineering department), Nabeel Ayoub (the ion source mechanical designer) and Yahya Nussaief (the head of the mechanical manufacturing workshops).

Despite having made that decision to switch to the calutron, it came to Dhafir's attention several months later that although Yahya Nussaief (head of the manufacturing activity in Directorate 3000) has indeed fabricated the calutron with more than one version to cater for one unknown dimension, yet they were left unemployed on the shelves of the workshop.

Those concerned with this affair had therefore to endure yet another heated debate over the issue to force through the use of the calutron, which was finally utilized. It eventually did produce a noticeable improvement to the output yield of the EMIS. The performance convinced everybody. It resulted in the final adoption of the calutron and to its replacement of the PIG ion source; after losing years of undue experimentation with the unfit choice of the PIG. Thereafter, when the industrial EMIS stage was implemented at the Al-Tarmiya site, the calutron ion source was chosen without dispute along with other specific design modifications from the Research and Development phase.

There was a further noteworthy incident regarding the use of the calutron. On one evening, the late Basel Al-Qaisi and Dr. Qais Numan came to Dhafir's office grumbling about the manner with which the operators of EMIS were handling the operation of the newly adopted calutron. They complained that the appropriate process parameters were not being implemented and had thus resulted in a reduced and fluctuating output. Dhafir accompanied both gentlemen to building 405 where the EMIS R 100s were being operated and asked the operator of one of the R 100 to step aside and allow Dr. Qais to operate it instead. Dr. Qais applied process parameters that were in-house modifications of those originally employed in the Manhattan Project as documented in the TID reports. Once he finished manipulating the controls to achieve the required parameters' levels, an immediate and significant increase in the output materialized. This incident was later on further refined by the operators themselves and even higher output figures were achieved with more stability.

The latest encouraging results of the use of the calutron led Dr. Jafar to call upon Hussain Kamel to witness and appraise the progress of the R 100s.

A meeting that was held in building 405 at Al-Tuwaitha and attended by Hussain Kamel, Dr. Amer Al-Saadi (who was the head of the Military Industrialization Corporation (MIC) at the time), in addition to Dr. Jafar, Dhafir Selbi, and Dr. Khalouk Refay, Dr. Thamer Numan and Dr. Moayad Maayof, the latter three were in charge of the EMIS research and development activities (referred to as activities 2A and 2B). However, the prepared demonstration by Dr. Jafar and Dr. Moayyed Maayof did not result in the expected positive appraisal awaited from Hussain Kamel. The main reason for the disappointment was due to the pointing out by one of the attendees of an overinflating of the results; explicitly, the presentation tried to compare old average outputs with the new maximum outputs, and not compare a previous average with the new average value.

Now that the calutron had become the main focus of the rejuvenated EMIS project, a newly created Information Activity within Group 3 headed by Dhafir in Directorate 3000 was established by the end of 1987 and headed by Dr. Imad Khadduri. Its main task was to find and disseminate all available data on the design of the calutron and its related EMIS process. The head of Group 3 had felt that due to the lingering reluctance of employing the calutron for so long, there was an urgent need to carry out a fresh thorough literature survey on the issue and to relieve the over tasked scientists and engineers from this chore. The intensive search of the few members of this activity in the library of the IAEC revealed that about 96% of the published literature on the caluton was actually already available in the library archives.

The main missing information on the calutron was the 164 patents pertaining to EMIS that were registered by the scientists and the engineers of the Manhattan Project during 1948 and 1949. All of the patents were easily purchased from the World Intellectual Property Organization (WIPO) in Geneva. They were catalogued and grouped according to their area of specialization and widely copied and distributed to the concerned scientists and engineers, each according to their area of interest.

The archived library records also indicated the presence in the IAEC library of the eighteen books that were published by the American National Nuclear Energy Series in the early nineteen fifties on the Manhattan Project. These books formed part of the American nuclear library gift that was delivered to Iraq back in 1956. However, nobody was aware of or has seen these books. After a thorough search of rarely accessed rooms in the IAEC library, the books were found in one locked trunk that was covered by layers of dust.

In order to facilitate the search in the contents of these books for urgently required technical and scientific details, the Information Activity personnel undertook the effort to scan the pages of all the books and employ an optical character recognition (OCR) software to index the words in all of the eighteen books that would enable to find the required word in any page in all of the books. The indexed textual database was distributed among all scientists and engineers allowing them to search for their specialized words of interest in all of the books simultaneously. Hard copies of the eighteen books were also provided to each department, unit and workshop. Capitalizing on the vigor enhanced by the new atmosphere to search through these documents in further detail, it was announced that for every member of the concerned Activities who would come up with a relatively important finding in the documents that has been overlooked and not used before in our work and would lead to their viable implementation in the ongoing projects will be awarded the maximum financial award. Several novel ideas triggered by this detailed search did come through and accelerated the work.

The head of Group 3 also located the source to a comprehensive library of about 5000 microfiche films with indexed catalogues that cross reference any product to its suppliers, its approximate price as well as to the industrial standards that apply to the selected product. He instructed the Information Activity to subscribe to this library, to the tune of a quarter of a million dollars annually. The library also contained the complete sets of all American and many European industrial standards, as well as American military standards. This service greatly accelerated the covert purchasing of required items and equipment for the INNP from 1988 onwards. Later, the same library, relocated in the Ministry of Industry and Minerals, was made available for free to all Iraqi institutions and tremendously served the rehabilitation and reconstruction effort that was undertaken by Iraqi engineers in the rebuilding of the damaged installations, especially in the power, oil and communication sectors, after the 1991 war.

4-1-3 Marginalizing exceptional scientists

The IAEC had adopted the practice of promoting its scientists by evaluating their research findings. In one of the IAEC research papers evaluation meeting, it was revealed that an evaluation of a research paper issued by the International Centre for Theoretical Physics in Trieste, Italy was outstanding. This center was founded in 1964 by Dr. Abdus Salam (Nobel Laureate) and had operated under a tripartite agreement among the Italian Government and two United Nations Agencies, UNESCO and IAEA. Its mission is to foster advanced studies and research, especially in developing countries. While the name of the Centre reflects its beginnings, its activities today encompass most areas of physical sciences and related applications.

The evaluation by this Center of the work of one of our scientists, Dr. Mohammed Abdul Zahra Habib, caught one of the commissioner's attentions. It evaluated his research paper as excelling in mathematics as well as in physics **in leading edge frontier topics** that the paper had dealt with.

For such a scientist in Iraq which had no Nobel laureates (the Manhattan Project had 12 Nobel laureates) and facing a scientifically challenging task that we were grappling with, it would have been necessary to fully engage Dr. Habib in the main stream activities of the INNP since its start and not side track him to scientific matters that are not related directly to INNP main projects.

Disappointingly, and after several attempted interventions, attempts failed to involve Dr. Habib in the main activities of Directorate 3000. The matter of interceding on behalf of Dr. Habib itself created tense moments. In the opinion of several prominent people in the IAEC, Dr. Habib could have advanced scientific results if he were allowed to be intimately engaged in our main activities.

One of the commissioners tried to engage Dr. Habib to start working on the mathematical modeling of the ionization of the PIG Ion source in order to provide insights that might enable possible solutions for the persistent problems that lingered on for years with the experimentation approach with no end in nsight. However, that effort was not supported. Dr. Habib was only engaged in Group 4 efforts late in the INNP effective life.

It was a striking irony that on the one hand we were experimenting for years with the PIG to no avail and its mathematical modeling was avoided, while on the other hand we had a proven calutron that was constantly being disregarded. Many could not explain this unjustifiable approach to the type of ion source, neither scientifically nor logically.

This was an example of the marked outcome of the imposed taboos, scientific mismanagement and the under utilization of potential scientific talent in the INNP.

4-1-4 The lack of clear planning with a prioritized approach

It was also apparent that Directorate 3000 did not have a totally interwoven plan leading to a definite synchronized aim for all participating sections that would set the final target with defined timelines and critical junctures. The overall plan that was being followed resembled a mixture of disjointed projects that were loosely coupled. That state of affairs resulted, as previously mentioned, in less than half baked results that ultimately ended in heated debates and recriminations.

There was a definite need for a process that was interactive in nature and guided by dynamic planning. The term (planning dynamics) was used in addressing this issue.

The concept of planning dynamics was introduced as a main tool to plan for projects in a comprehensive and dynamically interwoven modality. A planning centre lead initially by Dhafir Selbi, Dr. Abdul Qader Ahmed and Dr. Moayad Maayof was established to formulate the master plan for the program. Its responsibility was then transferred to Ghayath Al-Hashimi. The new work plan, with clear timelines and distribution of responsibility among the various departments was collectively agreed upon. Its implementation was to be supervised by the steering committee of Directorate 3000 (called Committee 3000) that will be detailed in section 4-1-6.

The main feature of the planning dynamics was its issuance of changed task priorities, depending on the progress of work, at the beginning of each week. These prioritized tasks would be distributed down the work flow chain all the way down and even to the workshop machinist standing at the turning mill as deemed necessary so that all would be on the same page regarding the priority of the tasks that are assigned to them. The top priority task is to be implemented first, and the priority is renumbered down the tasks list. Hence, the synchronized and coordinated implementation of the assigned tasks led to a marked decrease in wasted products, in addition to lifting the spirits of those involved in the design and manufacturing processes.

That kind of a management plan with a changing set of prioritized tasks was issued every week. It brought into reality the sense of feeling of being interacting positively to urgently demanded tasks according to their changing priorities. It also fostered a strong sense of coordination. The improved performance became apparent through repeated visits to all levels of responsibilities. For the first time you can see laid out on the drawing board or on a lathe machine the design for the part that we had only decided several days ago to give it the P0 level (top level) of priority. This was in stark contrast to previous such visits where you would see a machinist still working on a part which had been abandoned and was to be cast aside as it was replaced by the design of a newly replaced part several weeks earlier.

When the concept of planning dynamics was first introduced to the steering committee of Directorate 3000, there were some doubters who believed that it would be next to impossible to implement. Yet, once it did succeed, it was embraced by all involved and added effort was readily put into it in order to advance it.

4-1-5 Stratified management

In order to describe the managerial behavior that was practiced in Directorate 3000 prior to the reconfiguration in 1987, and had lingered for some time thereafter as its momentum diminished, it would be useful to shed light on the nature of some of the meetings that were held at that period.

During 1988, several 'campaigns' were launched to focus upon and try to resolve urgent problems. Each campaign lasted for several weeks during which everybody related to the problem at hand had to be present continuously, day and night, to resolve certain technological problems or obstacles. The participants would sleep in their offices and would rarely leave the working premises to visit their families during the particular campaign period. A series of meetings were held during the second scientific campaign which was initially declared to expedite progress of the various projects. These meetings were attended by almost all of the heads of activities and would usually start at

2 p.m. for six days a week. The meeting would usually last for 5 or 6 hours, and until backs ached, to end up with discussing only one or two points of the set agenda. The rest of the time was spent discussing issues outside the stipulated agenda. These side discussions usually did not require the presence of the more than 20 persons of different disciplines as the matters being discussed would be related to a small operational incident in one of the EMIS separators or other not so relevant matters. Discussions would get so diverse and fruitless to the point that many who attended began to label them as meetings that ended in negative results. We lost much valuable time in those meetings as well as bearing the suffering from our backaches with no rewarding advancement on most of the discussed points.

The topics of many of these meetings were soon relegated to the matter of discussing low level administrative affairs, while putting other leading staff members on a hold mode as they patiently waited for their chance to discuss technical issue, when and if that point in the agenda was reached.

At a point in time, just before the end of the Iran-Iraq war in 1988, and as part of the efforts to improve the efficiency of these meetings triggered by the new spirit, Dhafir and Dr. Jafar agreed, with Dr. Humam's consent, that Dhafir would chair of the second scientific campaign meetings in order to stem the diversification of discussions of the subject matters and abide by the agenda in order to focus on really urgent scientific and engineering matters.

As a side note, one of the extreme measures that were taken in order to goad discipline and to put the working staff in an accelerated tempo while working on strategic projects with tight time constraints was a decision taken by Hussain Kamel that any proven deliberate delay would expose the responsible to be taken to a court of law. However, this measure was never used.

4-1-6 The dissolution of a sacrosanct taboo

One of the important obstacles that faced Directorate 3000 was a sort of an opaque taboo that had engulfed its deliberations and had prevented any positive criticism to be voiced on the direction, conduct and planning of the scientific projects. This in turn led to further accumulation of simmering frustrations especially amongst leading scientists and engineers. Some of them did venture to step up and vent their frustrations to Dhafir right from the first day that he started working for Directorate 3000 in the spring of 1987. Many of their concerns were very revealing and startling to him.

Assisting Directorate 3000 in the spring of 1987 occurred at a precise time when the INNP was facing impeding difficulties and was saddled by a backtracked promise; this presented an opportunity in itself that allowed for the breaching of that taboo. It also allowed for the holding of open discussions with those who were immediately concerned about what should be changed, and how and when to change. As the news of this new openness spread, so had the outpouring of suppressed frustrations become apparent to all.

Upon digesting the aforementioned factors and their negative consequences of the performances of the projects, novel ideas poured forth which were widely encouraged by financial incentives.

One of the main organizational solutions that was proposed, and was finally adopted, was to create an operational entity called ZUMRA, which can be translated to a cell of people or a sort of a mini think tank in which each member was a representative of his specialized department and would report to it the issues discussed and bring back the required technical solutions.

The ZUMRA, for each required important constituent or system design, would include selected personnel, representing each and every one of the involved disciplines, who would collectively discuss and share their inputs regarding the design and fabrication of tasks during the same time slot. The resulting preliminary designs would be crafted interactively by those assigned to the discipline closest to the function of the required designed constituent or system, where usually a preliminary presentation is made which kicks off all other inputs.

The concept of the ZUMRA gave a visible uplifting push to the productivity of all concerned since it resulted in the desired designs and manufactured components that represented the collective inputs and views of all the related disciplines that were interactively arrived at and within a planned time frame.

This solution did not necessarily mean that we had a successful design for each instance that a ZUMRA's deliberations had ended in producing the intended component. However, it did guarantee a product whose quality was far superior to what had been produced earlier. It also facilitated the coupling of loose ended activities from different projects. It further put an end to the fruitless accusations and the passing around of the buck in blaming others for a useless product. That bickering would sometimes linger on for months. The ZUMRA's function reflected the positive aspects of the interactivity of all involved thoughts and experiences transpiring from the disciplines. Some international companies, such as Parsons Corporation, adopted the use of this ingenious planning procedure in the mid-nineties of the last century, i.e. more than ten years after its implementation in the INNP.

In addition to the ZUMRA concept, several other procedural and organizational reconfigurations were presented and swiftly implemented.

The biggest achievement of this overhauling of the interactive thought processes, and the reform of the workflow, was the breaking of the previously overarching taboo that led to the stifling of many constructive contributions that were intended to improving the situation, or to any public criticism of the archaic work patterns.

These included the concept of an ACTIVITY to replace the formal Department, and the concept of a GROUP that oversaw different activities. These concepts were in addition to the already mentioned concept of dynamic planning with its weekly reprioritization of the tasks that were to be implemented. Another practical concept involved the establishing of a dedicated Specialized Information entity to provide focused scientific and engineering information. These and many other suggestions were all presented to the Commissioners. The Commissioners quickly approved these novel operational concepts in early summer of 1987. They were promptly implemented during the remainder of 1987.

One such issue was why relevant matters that were directly related to the scientific approach, as well as the conduct and the planning of the projects, were not more widely discussed and was in fact limited to a closed circle of people that did not even include some of the Commissioners. The discussion of such matters was strictly confined to only the same few people who were of similar thinking. This state of affairs that had lasted for about six years had, in fact, prevented the presentation and development of novel and differing ideas. The belief that the lifting of these taboos would in fact provide endless new novel ideas was nurtured and finally established. These tight taboo clamps, and other unfortunate administrative measures, severely restricted the Iraqi brilliance from surfacing widely at an earlier stage of the INNP. Brave in heart, brilliant scientists and engineers came forward with ideas and suggestions whose limit was the sky. Had these taboos been lifted much earlier on in the program, the status of the INNP would have been much more advanced. Even with the belated implementation of these corrective measures, they still did not result in notable availability and marked progress of all of the required technologies primarily due to the very constrained time span of a few years before the demise of the INNP in the 1991 war.

The lifting of the taboos would not have been possible had it not been for the April 1985 and April 1987 milestones of the missed 1990 target.

According to the new organizational reconfiguration, Directorate 3000, while still maintaining its nomenclature to the prying eyes of the IAEA, was morphed into the following three Groups:

- *Group 1* that was dedicated to the centrifuge enrichment method after an earlier work on the diffusion method. Initially, the responsibility of heading this group was alternatively offered to Dr. Numan Al-Niaimi, then to Dr. Abdul Qader Ahmed who were both members of the steering committee of the Directorate 3000. Both of them refused to take this role and assume its duties and had their reasons for declining the offer. As a final resort, the position was offered to Dr. Mahdi Obeidi who accepted it immediately.

- *Group 2* that was headed by Dr. Jafar Dhia Jafar and was assisted by two deputies, Dr. Numan Al-Niaimi and Dr. Abdul Qader Ahmed. This group was dedicated to the Electromagnetic Isotope Separation (EMIS) enrichment method and its related chemistry and chemical engineering activities.

- *Group 3* that was headed by Dhafir Selbi and was dedicated to the Mechanical and Electrical Manufacturing, Specialized Information, covert and above board procurement and general administration. At a later stage, the electrical, electronic and mechanical engineering design activities were relocated from Group 2 and added to this group.

A Steering Committee (named Committee 3000) for the three Groups was formed from the following:

1. Dr Jafar Dhia Jafar, as head of the committee (He was also an IAEC commissioner).
2. Dhafir Selbi (He was also an IAEC commissioner).
3. Dr Numan Al-Niaimi (The rank of a Director General).
4. Dr. Abdul Qader Ahmed (The rank of a Director General). He was a prime mover in the efforts to establish this committee.
5. Dr. Mahdi Obeidi (The rank of a Deputy Director General).

Although some of the aspects describing the managerial behaviors and shortcomings did linger on for the viable life time of the INNP after the 1987 reconfiguration, major changes did take place. These were partly due to the heated follow up reports from the presidential office and eventually from Hussain

Kamel after the INNP (renamed as Petrochemical Project 3 - PC3) was attached to him. However, the fact that major changes were implemented was mostly due to the increasing outcry of the leading personnel regarding the futility of previous work patterns.

4-2 Attaching Group 1 to Hussain Kamel

In mid-summer of 1987, and after the formation of the above Groups, the Presidency Office called the IAEC deputy chairman informing him that it has been decided that Group 1 was to be transferred in its entirety and to be under the command of Hussain Kamel. That news came as a big blow to the newly reorganized Directorate 3000.

The presidential decision had to be implemented as soon as possible. The most challenging and difficult aspect of implementing the order were the decisions on who among the personnel should be included and what facilities and equipment should be transferred along with Group 1.

There was no doubt at the time that Hussain Kamel, who had been rapidly upgraded in status, had had the upper hand in this transfer. The posited argument by Hussain Kamel was that Group 1 had to be an alternative self sufficient project in all of its disciplines and activities, in the sense that it would be a parallel competitor to Directorate 3000. It also reflected the dissatisfaction of the higher authorities with the progress of Directorate 3000.

Although several personnel of the just splintered Group 1 remained at the geographical site of Al-Tuwaitha, Dr. Mahdi Obeidi and the majority of his accompanying staff left Al-Tuwaitha for a site appointed by Hussain Kamel. The remaining scientific and engineering personnel of Directorate 3000 were stretched even thinner.

In any event, this splintering was frequently raised as an excuse by both the reduced Directorate 3000 as well as by Group 1 in justifying their positions. It was readily blamed as the reason whenever either of the two was faced with ensuing difficulties and failures. In other instances, such pretexts were also reverted to for the support of other ulterior motives that extended far beyond the real effect of this splintering.

The years 1987 and 1988 witnessed yet other organizational milestone events. These included, notably, the dissolution of the Iraqi Atomic Energy Commission (IAEC) and the forming of the Iraqi Atomic Energy Organization (IAEO) in late 1987, to be followed by the formation of Petro Chemical 3 (PC3) in 1988 and the flip flopping in the attachment of a newly formed Group 4 (see section 4-3).

After the attachment of Group 1 to Hussain Kamel in the summer of 1987, all sorts of speculations began to circulate amongst the leading personnel of the remaining two Groups on the new role of Dr. Obeidi. He was known for his repeated preference of his theme: "It is how one should present the work rather than on the merit of the work itself". The speculation centered on how he would portray the situation in Directorate 3000 to Hussain Kamel, now being in direct contact with him, and what the real repercussions of such revelations on the IAEC itself would be since Hussain Kamel had an immediate channel to and the listening ear of late president Saddam Hussain.

The general conclusion at the time was that Dr. Obeidi would be reporting a very dark picture of the progress in the EMIS program. It was anticipated that the negative impression that he would be offering would be farfetched and, as such, would not be portraying the real state of affairs. This somber outcome became quite evident when Hamed Hummadi, the president's personal secretary, upon the directives of the late president, launched an official investigation into the commissioners shortly after Dr. Obeidi's[12] departure from the IAEC, as will be detailed in the next section 4-3.

---

[12] After the cessation of the 1991 war activities and the arrival of the IAEA inspectors, several directives till 1995 were issued to the INNP scientists and engineers to submit any and all work documents in their possession to their superiors and finally to representatives of the Military Industrial Corporation or else a penalty of death would be applied. Apparently, after the first such directive in 1991, Dr. Mahdi Obeidi hid several key drawings and small components of the gas centrifuge device, provided by a German company, in his garden. He never revealed their presence till after the occupation, even though they were one of the few final items that the IAEA inspectors insisted on obtaining before taking a decision to lift the economic sanctions. He then approached American troops to inform them

4-3 *The dissolution of the IAEC and the formation of the IAEO*

Some members of Committee 3000 had reasons to believe that more than one source of information from within the IAEC had reached the late president during 1987 and 1988. These inside leaks had cast serious doubts on the progress of the INNP under the patronage of Directorate 3000. But since these sources were not privy to the April 1985 pledge of the "goal" having to be achieved by 1990, the criticisms were not portrayed to him within that context; i.e. they could not claim that the 1990 "goal" was not a viable date any more since they did not know of it as we are revealing it for the first time in this book. This matter was kept top secret as the authors and fellow colleagues (other than Dhafir Selbi) first knew of it only in the summer of 2009 when we started working on the draft of this book.

---

of his trove but he was shunned. In desperation, he approached the journalist Kurt Pitzer who contacted David Albright, the founder of the Institute for Science and International Security (ISIS). Albright had earlier harbored the other CIA supported defector, Dr. Khidir Hamza, in 1995. Obeidi, and his family, were whisked to Kuwait for six months of interrogation were then taken to the US. The reason for Obeidi's hiding of the documents in his garden for 12 years is suspicious, and one can only assume that Obeidi did so either (1) to garner favor if ever the program was resurrected or (2) to barter it later for safe conduit to the US, which proved to be the case.

As a side note, in early 1987, a small group of physicists from IAEC was assigned to Hussain Kamel to form the nucleus of a group tasked with designing and later the manufacturing of a nuclear device. The unexpected and seemingly inflated requirements in time and resources of this endeavor, as demanded by that group, induced the transfer of the group to the IAEC, named Group 4 in the fall of 1987.

During the fall of 1987, the following commissioners were called upon jointly for an investigative session by Hamed Hummadi, the personal secretary to the late president:

1- Dr. Humam Abdul Khaliq, deputy chairman of IAEC
2- Dr. Jafar Dhia Jafar, Head of Directorate 3000.
3- Dhafir Selbi, Head of Department 4000, and Head of Group 3 in Directorate 3000.
4- The late Dr. Khalid Saeed, Head of Group 4 in Directorate 3000.

The only other commissioner who was exempted from these deliberations was the late Dr. Rahim Al-Kittal who had attended the April 1985 meeting with the late president.

Right from the onset of these investigative sessions, the attendees' positions split in two groups. The first group comprised of:

- Dr. Humam Abdul Khaliq

- Dr. Jafar Dhia Jafar

- The late Dr. Khalid Saeed

This group held the position that the present progress of the INNP projects was consistent and that their performances were solidly anchored. Yet, their position avoided touching upon the subject matter on whether the projected estimated progress of these projects would lead to and comply with the 1990 promised deadline.

While the second group, which comprised of Dhafir Selbi solely, maintained that whatever progress was being made, it came about in fact in a haphazard way, especially with regard to the EMIS main constituents. Furthermore, the few results that were materializing were without viable reasons. Moreover, the causes for repeated failures were neither fully understood nor known. It was maintained that since the causes of the failures were not known, they could not therefore be mastered. The finding of **solutions for unknown causes was not in the practical realm**.

These two conflicting positions led to a clear stalemate in the attempt to prove the convictions of either party to the listening Hamed Hummadi. The reason was clearly due to the fact that the arbitrator himself had no scientific background on the scientific aspects of the matter which was beyond his grasp and comprehension.

It became abundantly clear that the conclusions that Hamed Hummadi would be conveying to the late president would have been based purely on a personnel knowledge level rather than on the technical arguments that were raised. To it, the gentlemen of the first party had an accumulated experience in the IAEC of more than 60 years, with qualifications in the relevant fields of physics, while Dhafir Selbi, a mechanical engineer, had been in the IAEC for only 11 years by then.

The decision to dissolve the Iraqi Atomic Energy Commission (IAEC) and to create the Iraqi Atomic Energy Organization (IAEO) came swiftly thereafter in late 1987.

The consequence of that organizational transformation affected even the top hierarchy of the IAEC. The new organization excluded Izzat Ibrahim, the ex-vice chairman of the Revolutionary Command Council, i.e. the second top man in the country, who had been until then the chairman of the IAEC.

The decision that was made by the late president near the end of 1987 in dissolving the Iraqi Atomic Energy Commission and forming the Iraqi Atomic Energy Organization entailed the following new appointments for the ex-commissioners:

- Dr. Humam Abdul Khaliq, president of the IAEO.

- Dr. Jafar Dhia Jafar, the vice president of the IAEO.

- Dhafir Selbi and the late Dr. Khalid Saeed as director generals in the organization.

- The late Dr. Rahim Al-Kittal was transferred to the Foreign Ministry where he was later appointed as Iraq's

Ambassador to Vienna, Austria, the home town for the International Atomic Energy Agency.

The new hierarchy failed to put an effective limit on the debate of the different points of views among the members of the executive committee of the IAEC due to the fact that the ex-commissioners, who were of equal status, had remained in their positions. Hence, not much else developed regarding the directions of the IAEO that could have emerged. Indeed, the next milestone proved to deliver a fatal blow to the IAEO, which had then recently been formed.

4-4   The formation of Petrochemical 3 under Hussain Kamel

Sometime in the fall of 1988, and after the formation of the IAEO, Dr. Humam Abdul Khaliq, as president of the IAEO, had written an extensive report to the late president. In this he analyzed the status of the INNP's projects while he attempted to pinpoint the shortfalls in the progress of the program and offer solutions on how the situation could be remedied. There was no mention in that report of the already reported failure to meet the 1990 deadline.

It was believed that the report was put as a defensive shield against all of the attacks that were made on the progress of the INNP projects. It would have also served as a viable excuse to justify a possible officially announced backtracking step from the 1990 promise.

According to the analysis of Dr. Humam in the report, the shortfalls were mostly related to activities attributed to the failure of responsibilities of several manufacturing organizations within the Military Industrialization Corporation (MIC) which were serving the INNP) as detailed in Annex 2 and which were under the direct command of Hussain Kamel.

This report was also read by Hussain Kamel, as it was referred to him by the late president.

It was therefore decided that another round of official inquiry was to be held. However, this time, it was to be headed by Ahmed Hussain, the Head of the presidency office at the time, and to be attended by the concerned MIC and IAEO top officials.

That meeting, which took place in October 1988, proved to be fatal for the IAEO.

The attendees were:

- The MIC team, headed by Hussain Kamel and comprised:

    - Lt. General Dr. Amer Al-Saadi

    - Lt. General Dr. Amer Al-Obaidi

    - Other officers who were director generals of the manufacturing establishments in the MIC.

- The IAEO team, headed by Dr. Humam Abdul Khaliq, comprised:

    - Dr. Jafar Dhia Jafar

    - Dhafir Selbi

Ahmed Hussain started the meeting but it was eventually directly taken over and chaired by Hussain Kamel who launched heated attacks that were directed mainly at the president of the IAEO.

During these frontal attacks on the president of the IAEO, Hussain Kamel often tried to bring Dr. Jafar and Dhafir to his side by frequently referring his comments to them and asking them for their opinion on certain matters. However, their responses failed to assist him in this endeavor.

In fact, there was no real help that could have been offered to the president of the IAEO that would have shielded him from the vociferous Kamel series of attacks for he had an armory of piled up inside information for accusations.

As far as Dhafir was concerned during this deliberation, he was only responsible for explaining the status of the progress of the work on the construction infrastructure for the two EMIS enrichment projects at the Al-Tarmiya and Al-Sherqat locations. On these two projects, Hussain Kamel had no axe to grind, since both Dhafir and Harith Abood, the latter being one of the outstanding civil engineers, had thoroughly prepared the project's stance and made sure that there were no loop holes left through which Kamel would attack.

However, Kamel adroitly made use of what was apparently inside information that was passed to him on the status of the progress of work on the actual EMIS activities.

After several hours of the acrimonious debate, Ahmed Hussain requested to end the meeting.

Almost all the attendees had already figured out the eventual result of that meeting. Hence, it was of no surprise when the IAEO members were informed the next day that ALL of INNP activities were to be transferred to Hussain Kamel's command leaving the IAEO with the conventional NRC declared activities to the IAEA.

The transfer of the INNP to Hussain Kamel's command was the last important milestone for the years 1987/1988. The following steps proceeded:

- The formation of the Petro-Chemical No. 3 (nick named Petro 3 or PC3) which on paper was an entity of the Ministry of Industry and Military Industrialization but in actuality was linked directly to the person of Hussain Kamel. PC3 was to be headed by Dr. Jafar Dhia Jafar who was also appointed as deputy to the Minister of Industry and Military Industrialization for the additional purpose of complementing the cover-up for PC3's actual affiliations.

- The official start of PC3 was to be on the 1st of January 1989, but in actual terms it started under that status during November 1988.

- PC3 encompassed all of the INNP activities except the previously relocated Group 1.

- The management of the two enrichment sites at Al-Tarmiya and Al-Sherqat were transferred from MIC to PC3 and Dhafir Selbi was appointed to be responsible for them in addition to his work as Group 3 leader.

- Several other departments for safety, engineering services, specialized transport, documentation, administration, infrastructure designs and projects' follow-up were also established.

Hussain Kamel wasted no time in commencing his direct command of PC3.

One night during November 1988, Hussain Kamel called Dhafir asking him if he would go to Al-Tuwaitha, which Dhafir had just left only an hour earlier, in order to accompany him in visiting building 405 where the main experiments on the EMIS were being performed. Hussain did point out that prior to calling Dhafir he had tried to call Dr. Jafar but he could not get hold of him.

The rendezvous point was the roundabout of the suspension bridge on the Risafa side of the Tigris river; i.e. only across the river from the presidential complex where Hussain Kamel resided. The timing of the call and the request to immediately go to Al-Tuwaitha gave Dhafir the impression that the request was coming from the late president Saddam Hussain to check on things before embarking on something important.

Upon arriving to the rendezvous place, Dhafir found Hussain Kamel waiting in his car. One of his bodyguards took his car and Dhafir switched to Hussain's car which he was driving himself. During the drive to Al-Tuwaitha, which took around 20 minutes, Dhafir and Hussain were alone in the car and Hussain asked Dhafir a few questions about what does he think of the progress of the work in the EMIS experiments. Dhafir was as candid as he was during the previous meeting with Hamed Hummadi. However, taking into consideration Hussain's humble scientific knowledge and background, Dhafir did gather that he had understood the status in general terms but did not comprehend the probable far-reaching consequences of such a state of affairs.

By the time they had almost ended the tour of building 405, Dr. Jafar had managed to arrive at the site. All three of them rode back in Hussain Kamel's car, again with Hussain Kamel driving, with Dr. Jafar seated in the front seat and Dhafir in the back seat. Dr. Jafar could not miss the fact that Hussain Kamel was not at all satisfied with what he had witnessed. It was also clear that Hussain Kamel's dissatisfaction was not similar to Dr. Humam's; i.e. Kamel's dismay was not something with which one could determine the consequences.

At that time, Dr. Jafar happened to have been planning a trip the next day to Mosul where the INNP had established the Al-Jazeera uranium conversion plant. In order to defuse Hussain Kamel's dismay, Dr. Jafar promised him that once he returned from that trip, he would leave all of the administrative chores to Dhafir in order to enable him to concentrate fully on the scientific issues and to push things faster. It was clear that Dr. Jafar was attempting to defuse the disappointment of Hussain Kamel by implicitly promising accelerated advancements.

Chapter 5: PC3 January 1989 – January 1991

The vigor of the newly established Petro-Chemical 3 (PC3), under the authoritative and pushy command of Hussain Kamel, significantly compounded by the newly established efficient lines of communication and prioritization of streamlined processes, propelled and accelerated the activities of the INNP on several fronts:

- The accomplishments of the EMIS method that had reached its production stage though not without serious difficulties which required focused effort to solve (see Annex 1 for the three stages – basic research and development, laboratory scale separators and the production stage - and the related project designations of the EMIS method's development).
- The newly initiated centrifuge enrichment method that was directly attached to Hussain Kamel.
- The newly formed Group 4 that was relegated to the design and manufacture of the bomb's "device".
- The intense work on the chemical (solvent extraction and ion-exchange) methods of uranium enrichment, and
- A last ditch chemical extraction of highly enriched uranium from available fresh reactor fuel rods. This activity was initiated after the advance of the Iraqi army into Kuwait.

5.1 The production stage of the EMIS method

The concerted efforts on the experimental EMIS set ups at Al-Tuwaitha during the two stages, the basic research and development and the laboratory scale separators till the end of 1988, had resulted in encouraging results that included the use of multi-sources in one separator (project 106) and the design of a multi-magnet separator that consisted of a line of rotationally symmetric vertical and parallel dipole magnets that were shaped as a truncated flat double cone capped by a half magnet at either end of the line of magnets with interconnecting return iron pieces that were laid underneath the magnets to close the magnetic flux (project 105). This design was complemented by the perfection of supporting auxiliary systems such as the vacuum system, the control of the temperature of the feed material (UCl4), the liners and the ion beams collecting pockets.

The Al-Jazeera site, near Mosul in the north of Iraq, was also completed in 1989. It produced the feed material (UCl$_4$) to the production scale separators. This feed material was converted from uranium dioxide that in turn was extracted from the natural uranium yellow cake that resulted as a byproduct of the phosphate fertilizer production process at Al-Qaim facility, near the Syrian border and transported by rail to Al-Jazeera site.

The chemical research that was directly related to the EMIS project had equally achieved substantial leaps in the chemical treatment of the ion source holder, the liners and the extraction of the enriched material from the graphite pockets by washing the pockets in nitric acid inside stainless steel tanks which resulted in uranium nitrite. Relatively sophisticated analytical laboratories were established in building 240 at Al-Tarmiya to handle all kinds of analyses required for EMIS activities.

The total output of all of the experimental separators in the first two stages of development (basic R&D and laboratory) was about 448 grams of enriched uranium with an average enrichment of 5% by the cessation of operations near the end of 1990 and before the start of the 1991 war.

Eventually, the separator designs were scaled up to the third stage, the production stage, and a program for the manufacture of the separators to be installed at the Al-Tarmiya site was rapidly implemented in early 1989 with active participation of the Military Industrial Corporation's establishments that are mentioned in Annex 2.

The plan was to produce two complete separators per month to be installed next to the operating ones. The final implementation plan was to have two lines of thirty five separators, for a total of seventy operational separators, each of 120 cm in diameter with four ion sources and employing four sets of pocket collectors with a designed ion current of 150 mA each. This assembly would have provided for the first phase of uranium enrichment reaching up to 18% and with a projected design production rate of 69Kg of enriched uranium per year.

The next stage, to increase the enrichment from 18% to 93%, was to be implemented in a second phase that entailed the use of the twenty separators each of 60cm in diameter, with two ion sources and employing two sets of pocket collectors with a designed ion current of 50mA. The second phase's designed production output rate was expected to be 13Kg of 93% enriched uranium per year.

By the onset of the 1991 war, the essential buildings at Al-Tarmiya site that housed the various aspects of the EMIS project and required services were finished and operating, and the structures of the alternative EMIS production site at Al-Sherqat were nearing completion.

When the 1991 war commenced in January 1991, only eight 120 cm diameter first phase production separators were installed and operated at Al-Tarmiya achieving a peak current of 120 mA (not average current) from each ion source, while seventeen others were manufactured and were ready to be installed. Only five out of the twenty planned 60 cm diameter separators for the second enrichment phase were manufactured but none were installed.

At the start of the 1991 war, the total output of the Al-Tarmiya operating separators was merely 685 grams of only 3% enriched uranium, which represented a marginal quantity and quality of the designed production rate for this stage.

5.2 The nascent centrifuge uranium enrichment method

Initial work on this method, which started in 1987, entailed the development of an oil-bearing gas centrifuge (for which extensive design information was available in the open literature) with the goal of a production capacity of 10kg of 93% enriched uranium per year by 1994.

This task was undertaken by the Engineering Design Center (EDC) that was formally designated as Group 1. From the onset, the team in EDC concluded that Iraq's existing manufacturing capabilities were insufficient to produce the rotating components of the centrifuge machines to the required accuracy and quality.

Hence, EDC subsequently approached several machine tool suppliers in Germany, Yugoslavia, and Switzerland. They also sought the assistance of the German firm H&H Metalform, which established a liaison with a former MAN Technologie AG employee. He in turn cooperated with another ex-MAN Technologie employee in providing ECD with detailed design drawings, along with 170 technical reports, related to the production and operation of the centrifuges that were under development by the URENCO Group in the 1970s based on a carbon fiber composite rotor. They also provided for the supply of several trial rotors.

ECD also managed to obtain twenty-five pieces of maraging steel from an unidentified source, nineteen pieces of which were machined into centrifuge parts at the Nasser Engineering Establishment of the Military Industrial Corporation (MIC).

In the spring of 1990, the first magnetic centrifuge using a carbon fiber composite rotor was successfully assembled and tested at an operating speed of 60,000 rpm over a period of several months in a mechanical test bench.

Gaining confidence in the success of the gas centrifuge enrichment technology, ECD contracted with local and international organizations for the construction of the Al-Furat facility that was intended for the mass production of the centrifuges and for a pilot-scale cascade hall. The work on the facility was aborted with the start of the 1991 war.

5.3 The weaponization of the INNP[13]

As mentioned in section 4-3, and at the onset of Hussain Kamel's overt intrusion in the activities of the INNP, a small group of physicists from the IAEC was assigned to be directly under his authority in early 1987 to form the nucleus of a group tasked with the designing of an atomic weapon. The unexpected and seemingly exaggerated requirements in time and resources for this endeavor, as demanded by that group, induced the transfer

---

[13] WMD profiles: Nuclear, Iraq's Nuclear Weapon Program.
http://www.iraqwatch.org/profiles/nuclear.html

of the group to the IAEC and were named Group 4 in the INNP in the fall of 1987. The group was headed by the late Dr. Khalid Saeed and established its facility at the Al-Atheer site, 25 kilometers south west of Baghdad.

A special unit in the Al-Qaqaa establishment (that belonged to Kamel's MIC) was created to assist Group 4's scientists and engineers in the development and manufacturing of the high-explosive lenses and ultra-fast detonators that were required for the implosion device that would compress the highly enriched uranium core into a critical self-sustaining exploding mass.

The combined group developed several related manufacturing processes, such as rigid die-pressing of mixed explosives, plastic-bonded explosives, atmospheric and vacuum casting of melt-cast explosives and the casting of explosive polymer composites. By the end of 1990, they also managed to perform computer numerical controlled (CNC) machining of high explosives.

During 1990, the Al-Qaqaa team developed and produced plane wave lenses with various diameters (up to 120 mm) and various lengths. These lenses were tested and used as plane wave generators for shock-wave experiments.

Work on spherical lenses (to compress the uranium core ball) had started as early as 1988 with experiments using various kinds of explosives, including Baratol, PETN, COM-B, TNT, RDX and HMX. Tons of HMX were imported, and the team gained considerable experience in the casting of this material.

The combined team also mastered the design of dedicated exploding bridge wire (EBW) detonators, after experimenting with several types. In fact, the U.S. Departments of Defense and Energy permitted the attendance of three Iraqi scientists from Al-Qaqaa in 1989 at a quadrennial international detonation conference in Portland, Oregon, where nuclear weapon detonation technology and flyer plate technology were presented. The latter is used to control the force and shape of implosive shock waves.

By the end of 1990, Al-Atheer's site consisted of specially designed buildings intended to camouflage its real purpose (the only bomb that was dropped on the Al-Atheer site during the saturated bombing of Iraq in the early months of 1991 was a heat seeking bomb that targeted an electric substation outside its perimeter). The layout was designed for carrying out experiments on the highly explosive materials to about 1 ton with safeguarded rooms to prevent any atomic toxic material from escaping during the experiments on the high explosive lenses and detonators. The site also contained a building specially designed for the melting and casting of highly pure uranium metals.

The IAEA inspectors and Intelligence Agencies only learned of the Al-Atheer's significance after the lapse of five months from the beginning of their inspections inside Iraq. After realizing the importance of this establishment, the IAEA inspectors assembled Al-Atheer's leading scientists to witness the inspectors' demolition of the entire site with explosives in the fall of 1991. This uncalled for 'lesson' was another example of the malevolent intentions of those agencies that were steering the inspectors, as it would have been possible to blow up the premises without the need to assemble its scientific and engineering staff to witness the event. As such, the intent was clearly to humiliate and cower them.

5.4 Advanced chemical methods of uranium enrichment

The INNP achieved notable progress in the chemical (solvent extraction) and ion-exchange methods of uranium enrichment before the 1991 war.

The main purpose of undertaking the chemical enrichment method was for the provision of alternative slightly enriched uranium feed material for the EMIS separators instead of the use of natural uranium, thereby boosting efficiency and productivity of the EMIS separation.

Though successful research work in the solvent extraction process had been limited to basic research and development with laboratory-scale results, the concerned scientists had firm confidence that they would overcome any practical problems during the scaling-up of the process to a production stage. They initiated the procurement of the required components for a pilot plant that would have produced four metric tons per year of 1 to 1.2 % enriched uranium.

The initial experimental results of an ion exchange enrichment method also seemed promising. However, a project for the establishing of a pilot plant for the production of four metric tons per year of up to 3% enriched uranium had not gone beyond a preliminary assessment of the required equipment and materials.

The most promising project, which was still at the conceptual design stage in late 1990, combined both of the above chemical enrichment methods in a hybrid process employing for its first stage a solvent extraction method to be followed by an ion exchange output stage that would have produced up to 5 metric tons per year of 4 to 8% enriched uranium. This effort was abandoned due to the lack of human and material resources.

5.5 A last ditch chemical extraction effort for acquiring highly enriched uranium

In mid 1990, a crash program under Project 601 was launched to directly extract the highly enriched uranium that was contained in the fuel rods of Iraq's research reactors at Al-Tuwaitha to obtain about 41 kg of the highly enriched uranium 235 from the available supply of research reactor fuel from Russia and France that was already in Iraq's possession.

By December 1990, a chemical processing plant had been installed in the LAMA building[14] at Al-Tuwaitha with the intention to make available 26 kg of highly enriched uranium within 2-3 months. This building was severely damaged during the 1991 war (see picture referenced in the footnote).

---

[14] http://www-ns.iaea.org/projects/iraq/tuwaitha/lama.asp

Chapter 6: Shreds of the bombed INNP

The aftermath of the war on Iraq that commenced on January 17th 1991 and its demolishing of Iraq's National Nuclear Program followed by the meticulous dismantling of what had survived by the IAEA inspectors and until the occupation of Iraq in March 20th 2003 has been well documented by several Iraqi scientists as well as by the International Atomic Energy Agency's inspectors, as referenced in Chapter 1. Such information is also found among many other investigative reports after the occupation of Iraq in 2003, such as the shelved Iraq Survey Group's interim report by David Kay[15] in September 24, 2003 and by his successor Charles Duelfer[16] as well as belated IAEA's *mea culpa* declarations.

There are, nevertheless, three related aspects to this period that need to be highlighted:

- How much did the foreign Intelligence agencies in fact know of INNP's spectrum of activities?

---

[15] Report on Iraq WMD shelved as no evidence found by US-UK team http://www.informationclearinghouse.info/article4708.htm

[16] Council on Foreign Relations: Weapons of Mass Destruction and Iraq, May 24, 2005. http://www.cfr.org/publication.html?id=8157

- What was the role of the INNP management in the decision to keep the INNP from the IAEA's inspectors' eyes secret?

- What was the real intention of some of the IAEA inspectors on the ultimate fate of the INNP?

After the full exposure of the scope of the INNP, it became evident that foreign intelligence agencies were not able to learn of many aspects of the Iraqi National Nuclear Program, such as its covert procedures and bogus channels that were adopted for the purchase of sensitive equipment and materials and the pure fluke coincidence that lead to the bombing of the Tarmiya's Al-Safaa site which was the main establishment for the production facilities for the enrichment of uranium, as will be detailed in the section 6-1.

It also became apparent that in spite of the large sized establishment that employed more than 6,000 employees who were engaged in the INNP, none of the Intelligence Agencies had become aware of the wide scope and range of the Program's activities. This is itself a testimony to the loyalty of those who worked in the INNP and their commitment not to divulge any national secrets which is contrary to the adverse mass media insinuations that were leveled against Iraqis after the 2003 occupation.

After the cessation of war activities in 1991, the top management of the INNP sought to have a complete disclosure of the activities of the INNP to the IAEA inspectors. This was to ward off the adverse and bitter consequences that would befall the Iraqi people; however, their original advice was not heeded and it was only adopted after strenuous hide and seek encounters, as will be detailed in section 6-2.

The real intentions of some of the IAEA inspectors were far from being purely professional as their hidden agendas were laced with various intelligence gathering assignments. The chance was opportune for them, on behalf of foreign intelligence agencies, especially American and British, to investigate what was obscure about Iraq and its political structure and to survey the contents of sensitive establishments. In the final analysis, the persistently exaggerated and suspecting IAEA reports were the excuse that were relied upon to wage an already predetermined invasion of Iraqi in 2003. The IAEA inspectors had the regrettable support of Mohamed Al-Baradei, the Director General of the IAEA, in that he did not take a clear stand on the point that Iraq could not rejuvenate a shredded nuclear program and instead clung to three minor and irrelevant issues[17] in demanding a further few

---

[17] The three issues were extensively reviewed by Gary Dillon, the head of the IAEA inspectors, (see section 6-4) during 1998, including engineering designs that had not arrived to Iraq, several designs and a few small parts of the centrifuge machines that were hidden by Mahdi Al-Obeidi (see section 4-2, footnote 12) all of which would not constitute but a very minute portion for the supposed rejuvenation of the destroyed Iraqi nuclear program. Why, then, did Al-Baradei take a pretended hesitant stance and request for more time for inspections if not to purposely raise doubts about Iraq's capability to revive its nuclear program when these

months extension for the inspections. The U.S. exploited his pretended shaky position for an excuse to launch its occupation of Iraq.

Al-Baradei did not announce his regret for his actions until April 2011 when he tried to offload his own responsibility for the occupation of Iraq and dump it upon the shoulders of the Bush administration, going as far as clamoring for its legal accountability. However, this pang of conscious was only manifested during his campaign for the coming presidential elections in Egypt after several Egyptian revolutionary groups pointedly blamed him for his part in the destruction of Iraq. Many believe that Al-Baradei received the price for his dishonorable position on Iraq, which he is now trying to ameliorate, by accepting the Nobel peace prize money. We wonder just what kind of peace does Al-Baradei support?

6-1 What did we just bomb?

It is noteworthy to point out that the uranium enrichment facility at Al-Tarmiya (nicknamed Al-Safaa) was not amongst the targets of the intensive American carpet bombing campaign up until the 15th of February 1991, i.e. only 13 days before the end of hostilities.

---

very minor issues would in no way be significant for such a revival. Hence, what price did he receive aside from the Nobel Prize ?

This gives the impression that the relevance of the Al-Safaa site was unknown to the attackers, as well as to other declared and covert states in "the coalition of the willing", despite their combined intelligence agencies' coverage and intrusion. This suspicion was confirmed by the fact that the first attack on Al-Safaa included only three bombs dropped on the three largest buildings in the site. These were the production hall housing the 8 EMIS separators, the services building and the chemical processes building. The roof of the production hall completely collapsed with its two 80 ton cranes falling on the below separators and equipment. The extent of the destruction was revealed from the air. On the following day, on the 16th of February, a B52 bomber carpet combed the Al-Safaa site in a stretch extending beyond the entrance of the site. This was followed two days later by three more smart bombs. During these attacks, the authors and contributors of this book continued to be present on site working in shifts in order to manage small groups dealing with the emergency situations in order to prevent the loss of any personnel.

In 1992, IEEE spectrum[18] published a report quoting the pilot who dropped the first three smart bombs on the 15th of February in which he stated that as he was flying over the area, returning from a bombing mission with some bombs left, he had noticed a site with predominantly green color buildings by the

---

[18] "**Seeking nuclear safeguards. I. How Iraq reverse-engineered the bomb**", Zorpette, G., Spectrum, IEEE Spectrum, Apr 1992, Volume: 29, Issue: 4

Tigris river. He decided that he would randomly bomb the three biggest buildings. On routine photographing of the bombed places after such a mission, it was noticed that intense beehive activities were undertaken around one of the buildings (which was the production hall). This gave rise to the suspicion that it was indeed an important site and it was decided to carpet bomb it the next day using B52 bombers.

It was reasonable to assume that a prompt comparison of the layout of the Al-Safaa with other air picture images of large installations had revealed a strong similarity of the buildings and their lay out with that of another site, near the city of Al-Sharqat (nicknamed Al-Fajir) which was about 250 kilometers north of Baghdad. This was the alternative site for the EMIS enrichment facilities that was being completed at the time and at whose residence facilities, at a distance of a few kilometers, the families of the lead Iraqi team's scientists and managers were housed as a safe haven during the armed conflict. The Al-Fajir industrial complex was heavily bombed just a few days before the cessation of hostilities. Fortunately, its residence complex was not bombed, as had occurred earlier in February 1991 near the Samarra's Industrial facility in which 50 of the establishment's engineers' families (women and children) had perished due to the bombing of their residential site.

6-2 *Should we have played Hide and Seek?*

After the cessation of the 1991 war activities, the UN adopted resolution number 687 calling for IAEA inspectors to descend upon Iraq to uncover the full spectrum of the IAEC activities and programs.

Consequently, Hussain Kamel requested the INNP to submit a detailed list of all of its activities and sites to him for consideration on how to come to terms with the UN's resolution. As reported by Dr. Jafar and Dr. Al-Nuaimi's book "The Last Confession", page 148, the top INNP management met and put together the required information and maps. They submitted their report to Hussain Kamel with their recommendation that Iraq declares its INNP activities and open their sites for the IAEA inspectors in order to hasten the lifting of the economic sanctions and mitigate its debilitating effects on the Iraqi people.

In the decisive meeting between Hussain Kamel and the INNP top management that was held in April 1991, Hussain Kamel rejected the collective INNP's recommendation and ordered instead the conjuring up of alternative industrial projects and scenarios that would mask the true mission of the destroyed covert establishments. Thus began the hopeless whirlpool sequel between the IAEA inspectors and the shattered INNP that dragged on till October, 1997 when Iraq submitted its' Full, Final and Complete Disclosure (FFCD) report to the UN, including the full scope of its INNP activities.

6-3 *One of these scenarios*

In order to implement Hussain Kamel's above mentioned decision in concealing the nature and purpose of the Tarmiya's Al-Safaa site, and after removing all of the equipment that was present in the production hall, a thick layer of cement was poured over the return magnets as they were too heavy and time consuming to remove. This would make them appear to be production platforms. A scenario was agreed upon that maintained that this hall would be specialized in the production of high voltage transformers (400KV) as we had all of the related quotation documents for their purchase. This was indeed intended to be implemented in one of the establishments of the Ministry of Industry and Minerals. During the first meeting with the initial IAEA inspection team, which visited the site, they were presented with this scenario which they completely accepted and reported back: "the Al-Tarmiya site is not a nuclear site and has no connection with any nuclear activity". However, one of the workers at the site, who had just joined Al-Safaa three months before the start of the 1991 war after his return from his studies in the U.S., had escaped through the north of Iraq after the cessation of the fighting and returned to the U.S. exposing there the true nature of the Al-Safaa site at Al-Tarmiya. This led to a second visit by an IAEA inspection team that included four American inspectors with instruments that measured magnetic fields. Nonetheless, they did not find what they were looking for and returned empty handed.

Tellingly, the same person who insisted on hiding the INNP activities; namely, Hussain Kamel, took a surprising decision and ordered the handing over of all related Al-Safaa equipment to the army. This abrupt tack threw the entire hiding procedures into chaos with the resulting fumbling due to the nature of these pieces of equipment; namely, the very large 6 meters diameter magnets that weighed more than 60 tons each. Some of them fell to the side of the heavy army transporters on a public road from Al-Tarmiya to Baghdad where it was pounced upon by an IAEA inspection team.

After many heated discussions and interventions, a presidential order was issued in July 1991 to expose all of the INNP activities. That spelled the end of the hiding game and led to the destruction of whatever remained that had escaped the aerial bombardments. That also led to acrimonious friction with the IAEA inspectors. They were not performing in a professional manner due partly to interventions by intelligence agencies allegiances, such as David Kay who joined the IAEA inspector's team after the 1991 war. He publicly declared his CIA connection after the occupation of Iraq in 2003; he was also tasked by President Bush to lead an American team to "find" the weapons of mass destruction. He reported back in September 2003 that they had failed to find such a threat contradicting Bush's raison d'etre for the illegal occupation of Iraq.

On 25, April 1998, Iraq handed over to the IAEA its "Final, Full and Complete (FCCD)" report disclosing all aspects of its Iraqi National Nuclear Program.

6-4 An IAEA inspector's concern

In the autumn of 1998, with the UN inspections teams' excruciating prolongation of the nuclear file on Iraq, Dhafir Selbi (who had retired from the PC3 in September 1991) received a call from Dr. Jafar informing him that Gary Dillon, the Head of the IAEA inspection team at the time, was coming to Iraq to discuss three outstanding issues. One of these issues was related to a contact made by a Pakistani Journalist regarding an offer of a nuclear device design.

The issue was related to a proposal that was sent from the Iraqi General Directorate of Intelligence a few months before the start of the 1991 war to the INNP in which they claimed that a Pakistani journalist, who was assumed to have had close relationship with the Pakistani nuclear scientist Abdel Qadeer Khan, had approached them with an offer of providing the complete designs of the nuclear device, the activity that was the main target of Group 4's work.

Dillon had brought with him copies of documents that were garnered by David Kay from his raid on the Worker's Union Building in September 1991 that indicated the fact that there were distinct differing positions between Dr. Jafar and Dhafir regarding the response to this offer of assistance. He sought clarification of that matter in a session videotaped by the IAEA.

Dhafir had suggested that the offer should be accepted after veiling the request as not originating from Iraq and taking into consideration the lack of concrete progress in Group 4. Dr. Jafar had refused this suggestion, yet Dhafir went ahead with his proposition and informed the General Directorate of Intelligence in October 1990 to go through with the deal with the conditions attached. However, the Directorate never replied and nothing materialized. Dillon's piqued query was how did Dhafir come to discard Dr. Jafar's directive and had recommended to go ahead with the proposed deal? Dhafir's response was that what had mattered in the end, at that time, was achieving our target as promptly as possible, and that he was of the opinion that if the INNP can acquire such help then it should obtain it, especially when it was sought for Group 4's activity which was lagging.

Dillon was also attempting to find out whether this assistance had in fact been delivered to the INNP, and if so where would these designs be found.

It was obviously clear to us, from the belated date of focusing on this issue in 1998, as the dates of the correspondences on this matter between Dr. Jafar and Dhafir and between Dhafir and the

Iraqi General Directorate of Intelligence had occurred in the October and November of 1990 and that David Kay had obtained them since 1991, that Dillon was looking for a pretext to keep the Iraq nuclear file open in support of the continuity of the UN sanctions.

In any event, Dillon found it difficult to cement a pretext on this issue. Dhafir pointed out to him that the UN inspectors had the actual Group 4 designs and an expert would clearly discern whether these designs had incorporated any outsiders' designs, which they did not. Dillon's dismay was reflected in his final wicked question to Dhafir: "By now, all of the equipment and buildings related to your program have been destroyed. **What do you think we should do with the thousands of personnel involved in these projects**?"

Dhafir's immediate response was: "**Do you want us to ask them to commit suicide**"? Dillon blushed as he fumbled for a response to retract from his insidious question.

Conclusion

How far did the Iraqi National Nuclear Program reach before its demise?

There have been many estimates brandished about as to the time interval that was necessary for Iraq to attain a nuclear bomb capability in the event the 1991 war had not occurred. Some estimates had posited a matter of a few months while others claimed that it was only a matter of one year, or two years, etc.

We shall attempt to estimate the probable time period required relying on veritable and evidence-based analysis. What will follow is an extrapolation of the estimated required time taking into consideration the experience of the authors and other colleagues who worked in the INNP. In any event, the attempt remains within the realm of the estimation, subject to variances that accompany any estimated factor, in addition to the complexity of the events in the region and the challenging difficulties of the aim itself.

We will be focusing on the main parameters that were considered at the time as being the most crucial parameters in the critical path leading to the final target. Other parameters, though related to the progress of the INNP, are not discussed here since they were considered at the time as being easily manageable.

By the end of 1990, the status of the installation of the separators had progressed to the manufacturing of 17 large production separators designed to achieve 18% uranium enrichment levels[19], and the manufacturing of 5 smaller sized

---

[19] During the operation of one of these large separators (number 9), the achieved enrichment was 15% which confirmed the potential of these large separators to deliver the aimed at enrichment levels. (From the operations

separators designed to increase the enrichment of uranium from 18% to the required 93%. However, the number of installed and operational large separators was only 8 with the commencement of the 1991 military operations while none of the smaller separators were installed by then.

The scientists and engineers had focused on operating the installed large separators and had conducted several experiments aimed at attempting to reach a relatively high operational efficiency by obtaining the designed collected current value of 600 mA for a single separator. However, they did not manage to reach that operational current level before the start of hostilities.

The overall manufacturing, installation and operation plan for the separators called for the installation of two large separators per month. If we take into consideration that the target number of large separators had aimed to install 70 large separators and 20 small separators which would have called for the manufacturing and installation of the remaining 62 large separators that would have taken at least 31 months. We will also suppose that the concurrent installation of the small separators would have taken a year well within the above mentioned time frame.

Before we delve further into estimating the supposed time period that would have been required to obtain the bomb had

---

log book on 26-12-1990 written by Dr. Abdul Sattar Abdul Karim Al-Zubaidi).

the 1991 war not occurred, we must, again, clearly point out that the following time estimates are purely speculative and are open for divergent opinions. It is based upon personal experiences and interpretations of the concerned party and his organizational position within the total frame of the design, the manufacturing and the operational processes.

The basic parameters (**and in brief**) that would have determined the required period to manufacture the bomb are:

1- The operational procedures that were necessary to achieve the aimed for design values. The crucial factor in this endeavor was the improvement of the operational procedures in order to achieve the designed values after having taken the crucial decision on the type of the ion source (the calutron). Criticisms that had been asking for the concurrent conducting of experimental tests on the separators along with carrying out mathematical modeling of the ionization process in the ion source were beginning to sink in. In our opinion, had this parallel effort been adopted, it would have been possible to achieve the operational design values for the separators in a time frame of 16-24 months. It would have then taken a period of 10 months after that to be able to obtain the required highly enriched uranium that would be sufficient for one atomic bomb, taking into consideration the compiled quantities from both stages prior to achieving the designed production levels. This assumption excludes the fast track extraction of U_{235} from the available French and Russian fuel elements discussed earlier on.

2- The second important parameter was the program and the steps adopted to finalize the designs and the manufacture the bomb itself, without the uranium charge. This matter was facing more difficult problems relative to those in the uranium enrichment effort. Purely from conjecturing, we believe that had there been an infusion of new brains into this effort and correct practical and scientific approaches had been adopted to the problems faced by this program, it might have been possible to achieve positive results that would have been reliable within 36-48 months.

3- The other supportive factors, whether they were scientific, engineering, manufacturing or importing efforts, and despite the tremendous challenges that were faced or would have faced if the war had not taken place, would have been achievable within the upper time frames mentioned in points (1) and (2) above; i.e. within 48 months.

As a conclusion that is based on the above, we postulate that the Iraqi National Nuclear Program would have been able to manufacture one atomic bomb within 5-6 years from the end of 1990. This conclusion further takes into consideration the creative scientific and engineering capabilities of the personnel involved in the INNP and its achieved progress at the end of

1990. This conclusion also assumes that the separators' design values would have been achieved and that the manufacturing steps for the bomb would not be met by unseen difficulties.

However, the Iraqi National Nuclear Program was completely destroyed during and shortly after the 1991 military invasion. Yet, we maintain the genuine Iraqi spirit was not and will not be destroyed. The contents of this book gives testimony to the fact that what was achieved by the Iraqis in this program, while enduring extremely difficult circumstances that coincided with a fierce war waged during the duration of the program as well as a strangling embargo that blocked scientific and engineering exchanges with the world at large (even prior to the 1991 war), should be considered as an exemplary achievement. The Iraqis will continue to excel drawing from the fact that their civilization was the first in the world.

Annex 1: Summary of the stages of the EMIS program in the Iraqi National Nuclear Program

Iraq's most developed uranium enrichment technology focused on the electromagnetic isotope separation (EMIS) method. Its principle is similar to that of a mass spectrometer. An ion source produces plasma of the required compounds by electron bombardment. These ions are then accelerated by an electrostatic field and extracted as an ion beam that is then deflected in a magnetic field. The ions describe circular orbits with different radii according to their mass. The ion currents are magnetically focussed and deposited in an ion collector and their elements extracted.

The INNP carried out intensive scientific research and development activities on the EMIS at both Al-Tuwaitha and at Al-Tarmiya sites. It implemented the EMIS program in three overlapping stages.

1. Stage one: Basic Research and Development

This stage lasted from 1982 to 1987. It involved basic Research & Development of all aspects of EMIS. During this stage, the following projects were realized:

Project 101: Magnetic field project
Objective:

- To prove the agreement of the actual magnetic field configuration with the theoretical calculation. A closed type electromagnet was constructed with a mean radius of about 400 mm with relatively small gap distance.

Project 102:   Ion source with uranium tetrachloride (UCl_4) feed material
Objectives:

- To study the functioning and performance of the:

  - Magnetic field systems in the presence of the PIG type ion source in a 400 mm radius analysing electromagnet separator.
  - Ion source systems with noble gas and with UCl_4 vapour in the presence of relatively high magnetic field strength, which is needed for separation.
  - Ion current detection system, the collector system and the recording of the ion beam spectrum and measurement of ion currents.
  - Vacuum system.
  - Collector and liner treatment after operation.

Project 103:   Ion source with uranium hexafluoride (UF_6) feed material
Objectives:

It was originally planned to use UF_6 as the feed material. However, since this type of feed material was not available at the time, the decision was made to revert to UCl_4. This project was operated in a similar manner to Project 102.

2. Stage two: Laboratory scale separators

The second stage, which started in 1983, reached an experimental stage in 1987 and continued until the 1991 war. It concerned the stability of the separator, the collector design, recovery improvement and to achieve better availability.

Project 104

This project involved experimenting with magnets of 1000mm and 500mm radii in order to test the prototypes of the production scale magnets.

- ### R50 Separator

    The R50 represents the beta phase of the calutron ion source design. However, the tests that were made used natural uranium feed.

    ### Objectives:

    The objectives of this project were to perform:

- Studies on the behaviour of both the new magnetic field and the new ion source parameters
- Improvements on project 102 with the new dimensions.
- Studies on ion beam stabilization.
- Collector/Pocket investigations.
- Enrichment studies.

• The R 100/1 separator

This was the second separator in project 104 to come into operation. The mean radius of the magnet was 1000 mm.

The magnetic field strength B of the magnet was relatively lower than that of the R50.

Objectives:

The functioning and performance of the:

- New magnetic field systems in the presence of the large ion source.
- Relatively large ion source systems with UCl_4 vapour assembly.
- Water cooled ion current detection system.
- Vacuum system.
- Collector and liner processing after operation.
- Availability studies.

- **The R100/2 Separator**

  This was the third separator of the project 104 to come into operation. The mean radius of the magnet was 1000 mm and was the same as for Project 100/1.

  A different group conducted the experiments on this separator in order to independently correlate the results obtained from the previous one for Project 100/1. The results of the collected ion current, as well as the enrichment and the availability were similar.

- **The R100/3 Separator**

  This was the fourth separator of project 104 to come into operation. The mean radius of the Magnet was also 1000 mm.

  ### Objectives:

  To prove the concept of multiple ion sources in the same magnetic field. These were the same objectives as separators R100/1 and R100/2. However, it contained two ion sources and four pockets

During the last 12 months of operation in Al-Tuwaitha, the three R100 separators produced about 360 grams of enriched uranium with an average enrichment of about 7 percent. In total, all separators of Project 104 had produced about 640 grams of enriched uranium with an average enrichment of 7.2 % from the spring of 1988 through 1990.

Project 105

This was a scaled down model unit for the third stage of the EMIS program, i.e. the production stage.

Objectives:

- To investigate the setup of a multi-magnet series as an analytical tool for the magnetic field design.

- To examine the scaled down model of a production unit in a ratio of 1:5.

• Project 106

The aim of project 106 was to study the operation of a multi ion sources configuration and to investigate the possibility of the cleaning of the liners using the electrical discharge method.

3. Stage three: The production stage

The first industrial scale EMIS facility was constructed in Al-Tarmiya, 40 kilometres northwest of Baghdad. A second facility was planned at Al-Sharqat, 200 kilometres northwest of

Baghdad. The two plants were to be identical with a total of 70 first phase R120 separators and 20 second phase R60 separators in each. The 70 R120 separators (1200 mm ion beam mean radius) were to be installed in two large parallel piers with 35 separators in each line.

The initial plan was for installation of the first half of the separators at Al-Tarmiya, followed by the installation of the first half of those at Al-Sharqat. Then the second half would be installed at Al Tarmiya before completing installation at Al-Sharqat. At the beginning of 1990, the plan was changed to implement a full installation of the separators banks at Al-Tarmiya and then proceed with the full installation at Al-Sharqat.

The installation of the separators in the experimental hall of building 80 at Al-Tarmiya was to occur in several phases, the first eight R120 separators were installed between February 1990 and September 1990.

The R60 separators (600 mm ion beam mean radius) were to be installed in parallel in four phases of five separators each in experimental hall 90. The production of the first five seperators was completed, but none were installed by January 1991.

Annex 2: The engineering infrastructure that supported the INNP

The engineering concerted effort to support the INNP utilized some of the already available facilities at Al-Tuwaitha as well as the construction of new facilities in and around Baghdad for the production of the large amounts of equipment, materials and assemblies as well as facilities for the installation, testing and maintenance of the assembled units.

These are the major engineering facilities that supported the INNP:

1- The Al-Rabee' mechanical establishment (at Al-Zafaranya) which was equipped with sophisticated and specialized CNC machines, computerized lathe machines and grinding tools for the production of some of the EMIS separators and the ion sources.

2- The Al-Dijla electrical and electronics manufacturing establishment (at Al-Zafaranya) which was built for the design, manufacture and testing of the power supplies that were required by the prototypes and production separators as well as the control modules and racks used to control the cluster of separators. This workshop was furnished with facilities to produce electrical cabinets for different purposes, a system for designing and drawing of printed multilayered circuit boards and all other required electrical and electronic systems.

3- The Al-Amin (at Al-Yousifia), Al-Radhwan (at Abu Ghraib), and Al-Ameer (at Al-Ameria) mechanical manufacturing facilities to fabricate the large pieces of the magnets, the return irons, the vacuum chambers and various parts of the ion source and collectors assembly. These were in addition to employing the workshops of other Military Industrialization Commission (MIC) such as Bader and Uqba bin Nafia' (at Yousifia).

4- The Al-Jazeera site (code named the Wax Factory) near Mosul for the supply of the UCl_4 feed material in packs of 2 kilograms each for the production-scale separators at Al-Safaa. UCl_4 is produced from UO_2 which was in turn produced at this facility from purifying the raw uranium (yellow cake) transported by rail from the Akashat phosphate complex near the Syrian border. A UCl_4 preparation facility was also set up at Al-Tuwaitha that was used as feeds to the ion sources during the research and development stage of the prototype separators.

5- The Al-Tarmiya site (also known as Al-Safaa) was built to house both stages of the production-scale separators with all the supporting chemical labs for extracting the collected material in the receivers. It also provided all services required by the separators such as demineralized water, clean pressurized air, etc.. Only 8 separators were installed at the start of the 1991 war. The maintenance engineers stood by day and night to ensure the continuous operation of the separators with mean time between failures reaching as little as 3.5% so as to achieve the designed availability figure of 55% .

6- The Al-Sherqat site (also known as Al-Fajir) was built to be a replica of the Al-Tarmiya site for two reasons: one is for security reasons in that if either of the two sites covert status is exposed or is physically damaged, then the other would continue functioning. The second reason was that if an accelerated production was required, then there would be an alternative production site. However the Al-Fajir site was only 85% complete at the onset of the 1991 war.

7- Large quantities of liquid nitrogen were required for the vacuum systems. Hence, two facilities were built each with two lines of production capacity of 300 liters per hour. The first, Al-Amal, was built about 6 kilometers from the Al-Safaa site and the other was built within Baiji's oil refinery complex near Al-Fajir site. These two factories proved to be useful as the surplus production was supplied to other establishments.

Index

129

130

www.ingramcontent.com/pod-product-compliance
Lightning Source LLC
Chambersburg PA
CBHW070149290526
45789CB00002B/695